TOTAL
WAR

ALSO BY THOMAS POWERS

Thinking About the Next War (Alfred A. Knopf, 1982)

The Man Who Kept the Secrets: Richard Helms and the CIA (Alfred A. Knopf, 1979)

The War at Home: Vietnam and the American People (Grossman/Viking, 1973)

Diana: The Making of a Terrorist (Houghton-Mifflin, 1971)

ALSO BY RUTHVEN TREMAIN

The Animals' Who's Who (Charles Scribner's Sons/Routledge and Kegan Paul, 1982)

Teapot, Switcheroo, and Other Silly Word Games (Greenwillow, 1979)

The 1978 (and 1979) *Calendar for Children* (Doubleday, 1977, 1978)

Fooling Around with Words (Greenwillow, 1976)

Macmillan, Inc., Report on the Year for Young People (1973)

My Friends: A Self-Portrait Autograph Book (Macmillan, 1971)

Summer Diary (Macmillan, 1970)

Secrets with Ciphers and Codes, with Joel Rothman (Macmillan, 1969)

The 1964 Fortnightly Chin Up Calendar: A Comforting Collection of Terrible Things That Have Happened to Other People (Macmillan, 1963)

The 1964 (and 1966–1977) *Calendar for Children* (Macmillan, 1963, 1965–1976)

TOTAL
WAR

WHAT IT
IS, HOW
IT GOT
THAT WAY

**THOMAS
POWERS**

AND

**RUTHVEN
TREMAIN**

WILLIAM MORROW AND COMPANY, INC.
NEW YORK

LIBRARY OF CONGRESS

Library of Congress Cataloging-in-Publication Data

Total war : what it is, how it got that way / Thomas Powers and Ruthven Tremain.
    p.   cm.
    Bibliography: p.
    Includes index.
    ISBN 0-688-06919-3 (alk. paper)
    1. War—Quotations, maxims, etc. I. Powers, Thomas, 1940 Dec. 12– II. Tremain, Ruthven.
U21.2.T68   1988   87-29951
355'.02—dc19   CIP

Printed in the United States of America

First Edition

1 2 3 4 5 6 7 8 9 10

BOOK DESIGN BY MANUELA PAUL

CONTENTS

INTRODUCTION

This is a book about war. There have been lots of wars during the last forty years, but no big ones since 1945. There are many possible explanations for this happy fact. Some scholars of war say "nuclear deterrence" is the reason—all-out wars between Great Powers are just too dangerous now. Other scholars say a better explanation can be found in the exhaustion of Europe by the two big wars in the first half of the twentieth century—the wars that taught us what "total war" is like.

My own feeling is that war and peace are both mysteries. Every war has its causes, sometimes a whole list of causes, but while the causes tend to be highly particular, the wars are much the same. They are hard to get going, then even harder to stop. The original goals begin to look paltry, and soon cease to matter. The leaders of both sides dig in their heels. The generals learn to ignore the sight of blood, or lose their commands to others less squeamish. The soldiers and common citizens endure what they can and die when they must. It begins to seem the war will never end, but eventually one side falters, the balance of forces rapidly begins to tip, and a peace agreement of sorts is dictated by the victor.

Once upon a time the wars of Europe might end after a battle or two, but no longer. The style changed with the French Revolution of 1789. Since then wars have tended to end with big winners and big losers. Defeat hurts, and one origin of the total war of modern times can be found in the fear of total defeats like those suffered by the French in 1815, 1870, and 1940, or by the Germans in 1918 and 1945. Someone once said, "I've been rich and I've been poor, and believe me—rich is better." The same goes for winning wars. It's not what winners get, but what they escape, that makes the difference.

War itself has changed during the last two hundred years. The French Revolution introduced conscription on a mass scale and the mobilization of whole societies to fight. Traditional armies of a few tens or scores of thousands suddenly ballooned to hundreds of thousands, with casualties on the battlefield to match. Technical innovation made weapons more accurate and powerful, and industrial expansion produced them in thousands. The horrors experienced by the Prussian military writer Carl von Clausewitz in the wars with Napoleon gave him a glimpse of something he called "absolute violence"—firepower so deadly nothing could survive on the battlefield. Clausewitz described "absolute violence" as the logical goal of military operations, but at the same time dismissed it as unattainable. The "friction" of war—the sheer cussedness of men and things—would always limit violence to tolerable levels, Clausewitz felt.

But Clausewitz reckoned without the full fruits of human genius, once it had turned itself to military invention.

Introduction

During World War I, a century after the defeat of Napoleon, the Western Front in France was the scene of something very close to "absolute violence." Huge armies locked along a thousand miles of muddy trenches and pounded each other with artillery which turned the land to moonscape. When the Battle of the Somme commenced on July 1, 1916, the British army suffered sixty thousand casualties in a single day. Only twenty-five years later huge fleets of British and American bombers pounded German cities into rubble. They soon discovered the proper mix of ordnance—high explosive to blast open doors and windows and break up buildings into kindling, incendiaries to set the wreckage alight. When everything went right the result was a firestorm, a raging inferno that might level ten or fifteen square miles of city, and kill fifty or a hundred thousand people. This truly was "total war"—war without limits, moral, practical, or geographical. All was permitted, and all was possible. The introduction of nuclear weapons in the last two weeks of the war only made "absolute violence" easier and cheaper.

The "causes" of the last two big wars don't really explain very much; the causes are too small, and the wars too big. They are mysteries. Since we can't say what caused the wars, how can we say what caused the peace that has followed? Is it nothing more than fear? Generals and statesmen wept with fear before the First and Second World Wars, but we had them anyway. Is the fear somehow greater now? Twice burned, have we finally learned there is no escape from total war, once begun? I'm afraid I can think of only one way we might really answer these awkward questions, and that is to wait and see.

This is a book about war and what we have made of it

over the last twenty or thirty centuries. The short answer is that we have made it bigger and more lethal, and the pace of development has been especially brisk in the last few decades. The book is made up mostly of things people have said or written down in the heat of events. I collected them in the course of doing research for a history of strategic weapons. I am under no illusion they can be made to prove a case, but all the same, to me they suggest a way of interpreting the pattern of war so far, and what might happen next.

—THOMAS POWERS
South Royalton, Vermont
September 1987

It is a good thing for an uneducated man to read books of quotations.

Winston Churchill, *My Early Life* (1930). Compiling one works even better.

TOTAL
WAR

1

THE ATOM

The nightmare will not come true, and we shall go on living a life of reasonable happiness, committing follies and paying for them, and doing our duty in the old fashion, indifferent whether it is a horse we ride behind or a steam engine that wafts us on our way . . . even the atomic energy imagined by Mr. Wells would be powerless always against the unbroken traditions of the human race.

Blackwood's Magazine review of H. G. Wells's novel, *The World Set Free* (1914), in which he imagined the discovery of atomic energy, a period of unparalleled prosperity, and a war in 1958 using "atomic bombs" which left the major cities of Europe pools of bubbling radioactivity.[1]

[It takes one's breath away] to think of what might happen in a town, if the dormant energy of a single brick were to be set free, say in the form of an explosion. It would suffice to raze a city with a million inhabitants to the ground.

Hans Thirring, an Austrian physicist, in 1921.[2]

Total War

"You haven't lost anything if I don't discuss your work with you in detail. Its foolishness is evident at first glance."

> Albert Einstein in 1921 to a student who had suggested that Einstein's General Theory of Relativity could be used to make a nuclear bomb. Later, when told in the summer of 1939 that a chain reaction was possible, Einstein said, "That never occurred to me."[3]

Some physicists think that, within a few months, science will have produced for military use an explosive a million times more violent than dynamite. . . . And it is in America where the thing will in all probability be done, if it is done at all. The principle is fairly simple. . . . It is the old dream of the release of intra-atomic energy, suddenly made actual at a time when most scientists had long discarded it. . . . What will happen, if a new means of destruction, far more effective than any now existing, comes into our hands? . . . We have seen too much of human selfishness and frailty to pretend that men can be trusted with a new weapon of gigantic power. . . . Yet it must be made, if it really is a physical possibility. . . . Such an invention will never be kept secret, the physical principles are too obvious. . . . For a short time, perhaps, the U.S. Government may have this power entrusted to it, but soon after it will be in less civilized hands.

> C. P. Snow, *Discovery,* September 1939.[4]

"General Groves, you know—with respect to what you said to me—*my* reputation is already made. It is *your* reputation that depends on this project."

The Atom

The American physicist Ernest Lawrence, at lunch with General Leslie Groves, commander of the Manhattan Project, in October 1942. Earlier that morning, at Lawrence's laboratory at the University of California at Berkeley, Groves had said, "And so, Professor Lawrence, you'd better do a good job. Your reputation depends on it."[5]

We cannot afford after the war to face the future without this weapon and rely entirely on America, should Russia or some other power develop it.

Sir John Anderson to Churchill on July 21, 1943, explaining why Britain ought to build atomic bombs of her own.[6]

"I cannot see what you are talking about. After all this new bomb is just going to be bigger than our present bombs. It involves no difference in the principles of war. And as for any post-war problems there are none that cannot be amicably settled between me and my friend, President Roosevelt."

Prime Minister Winston Churchill to the Danish physicist Niels Bohr in a meeting on May 16, 1944. Bohr wanted Churchill's help to avoid a postwar arms race between the West and the Soviet Union.[7]

"It was terrible. He scolded us like two schoolboys!"

Physicist Niels Bohr, describing his meeting with Churchill.[8]

Total War

The President and I are much worried about Professor Bohr. How did he come into this business? He is a great advocate of publicity. He made an unauthorized disclosure to . . . Frankfurter who startled the President by telling him he knew all the details. He says he is in close correspondence with a Russian professor. . . . What is all this about? It seems to me Bohr ought to be confined or at any rate made to see that he is very near the edge of mortal crimes. . . . I did not like the man when you showed him to me, with his hair all over his head, at Downing Street.

> Churchill to his science advisor, Lord Cherwell, following Bohr's visit.[9]

It might perhaps, after mature consideration, be used against the Japanese, who should be warned that this bombardment will be repeated until they surrender.

> President Franklin D. Roosevelt and Prime Minister Churchill in an *aide-mémoire* after a meeting at Hyde Park, September 17, 1944.[10]

"Senator, you have no idea how right you are, but I think you'll discover before the year is over."

> President Roosevelt to Senator Arthur H. Vandenberg in March 1945, referring to a passage FDR had underlined in a Vandenberg speech which read, "If World War III unhappily arrives, it will open new laboratories of death too horrible to contemplate.[11]

The Atom

This matter now [of the atomic bomb] is taking up a good deal of my time and even then I am not doing it justice. It is approaching its ripening time and matters are getting very interesting and serious.

Secretary of War Henry Stimson, in his diary on March 8, 1945. [12]

As I heard these scientists and industrialists predict the destructive power of the weapon, I was thoroughly frightened.

Secretary of State James Byrnes in the spring of 1945, after hearing Oppenheimer say it would eventually be possible to build bombs with a power equivalent to one hundred million tons of TNT. No bomb of that yield has ever, in fact, been built, but in 1961 the Russians detonated a warhead equivalent to sixty million tons of TNT. [13]

"Now we are in for trouble."

Physicist J. Robert Oppenheimer, on his appointment to the Interim Committee studying the use of the atomic bomb, Spring 1945. [14]

We can propose no technical demonstration likely to bring an end to the war; we can see no acceptable alternative to direct military use.

Formal report to Secretary of War Henry Stimson rejecting proposals for a demonstration of the first atom bomb before actual use on a

21

target, submitted by Oppenheimer, Lawrence, Fermi, and Arthur Compton, the Scientific Panel of the Interim Committee, June 16, 1945.[15]

"Well, we're all sons of bitches now."

Los Alamos scientist Kenneth Bainbridge to Oppenheimer, on witnessing the first atomic bomb test.[16]

"If it explodes as I think it will, I will certainly have a hammer on those boys."

President Harry S Truman in July 1945, just before meeting with the Russians in Potsdam.[17]

It is certainly a good thing for the world that Hitler's crowd or Stalin's did not discover this atomic bomb. It seems to be the most terrible thing ever discovered, but it can be made the most useful.

Truman in his diary, July 1945, shortly after learning of the successful test of the first atomic bomb.[18]

"Now I know what happened to Truman yesterday. I couldn't understand it. When he got to the meeting after having read this report [about the successful atomic test at Alamogordo, New Mexico] he was a changed man. He told the Russians just where they got on and off and generally bossed the whole meeting."

> Churchill describing Truman at the Big Three meeting at Potsdam in July 1945.[19]

"We knew nothing whatever at that time [July 1945] about the genetic effects of an atomic explosion. I knew nothing about fall-out and all the rest of what emerged after Hiroshima. As far as I know, President Truman and Winston Churchill knew nothing of these things either, nor did Sir John Anderson, who co-ordinated research on our side. Whether the scientists directly concerned knew, or guessed, I do not know. But if they did, then so far as I am aware, they said nothing of it to those who had to make the decision."

> Former Prime Minister Clement Attlee in 1959, recalling the Potsdam Conference during which he became the British Prime Minister after Churchill resigned on July 26, 1945.[20]

"We'll have to talk it over with Kurchatov and get him to speed things up."

> Molotov to Stalin, after Truman told Stalin about the new "weapon of unusual destructive power" at the Potsdam Conference in July 1945. Igor Kurchatov was the Soviet Union's chief atomic scientist.[21]

Total War

I watched the assembly of this man-made meteor . . . and was among the small group of scientists and Army and Navy representatives privileged to be present at the ritual of its loading in the "Superfort." . . . It is a thing of beauty to behold, this "gadget." Into its design went millions of man-hours of what is without doubt the most concentrated intellectual effort in history. Never before had so much brainpower been focused on a single problem.

> William L. Laurence, reporting the bombings of Nagasaki in *The New York Times*, September 9, 1945.[22]

"My God, look at that sonofabitch go!"

> Captain Robert Lewis, co-pilot of the *Enola Gay*, overheard by fellow crew members as they all watched the mushroom cloud rise over Hiroshima. But Lewis silently wrote in his log of the mission, "My God, what have we done?"[23]

"The first thing I knew, there was a blinding white flash of light, and a wave of intense heat struck my cheek. This was odd, I thought, when in the next instant there was a tremendous blast. The force of it knocked me clean over but fortunately, it didn't hurt me. . . . Hundreds of injured people who were trying to escape to the hills passed our house. The sight of them was almost unbearable. Their faces and hands were burnt and swollen; and great sheets of skin had peeled away from their tissues to hang down like rags on a scarecrow. They moved like ants."

Dr. Tabuchi, reporting what happened to him in Hiroshima on August 6, 1945.[24]

Flames rose and the heat set currents of air in motion. Updrafts became so violent that sheets of zinc roofing were hurled aloft and released, humming and twirling in erratic flight. . . . A ball of fire whizzed by me, setting my clothes ablaze. They drenched me with water again. From then on I am confused as to what happened.

Dr. Michihiko Hachiya, recalling August 6, 1945 in Hiroshima.[25]

I saw a thin, woebegone dog making his way. . . . Most of his hair was gone, so I guessed he suffered radiation injury, too. Somehow, this dog was symbolic. What a dismal view.

Dr. Michihiko Hachiya, in his *Hiroshima Diary* entry for September 3, 1945.[26]

I was shocked and depressed beyond measure. The thought of the unspeakable misery of countless innocent women and children was something I could scarcely bear.

Total War

The German chemist Otto Hahn, discoverer of fission in 1938, describing his reaction to news of Hiroshima at the country house in England where he was interned with other German scientists active in nuclear research during the war. He was so agitated that night of August 6, 1945, that his colleagues watched over him till he fell asleep.[27]

"When I was a boy I wanted to do physics and watch the world make history. Well, I have done physics and I have seen the world make history. I will be able to say that to my dying day."

The physicist Max von Laue, one of ten German scientists interned at Farm Hall in Britain, late in the evening of August 6, 1945, after a long discussion of the bombing of Hiroshima with his colleagues.[28]

"The world will note that the first atomic bomb was dropped on Hiroshima, a military base. That was because we wished in this first attack to avoid, insofar as possible, the killing of civilians. But that attack is only a warning of things to come."

Statement by President Truman, August 9, 1945.[29]

"If we, a professedly Christian nation, feel morally free to use atomic energy in that way . . . the stage will be set for the sudden and final destruction of mankind."

Methodist Bishop G. Bromley Oxnam and John Foster Dulles, speaking for the Federal Council of Churches in August 1945, after Hiroshima and Nagasaki.[30]

"Everybody was moaning and wringing their hands."

Luis Alvarez, a physicist at Los Alamos during the war, describing the dramatic change in atmosphere after his return from Hiroshima. Elation at the achievement had completely disappeared.[31]

"Oppie says that the atomic bomb is so terrible a weapon that war is now impossible."

Popular report at Los Alamos after Hiroshima.[32]

"If you ask, 'Can we make them more terrible?' the answer is yes. If you ask: 'Can we make a lot of them?' the answer is yes."

Oppenheimer to *Time* magazine, October 29, 1945.[33]

"Mr. President, I have blood on my hands."

Oppenheimer to Truman in 1946.[34]

Total War

"Don't you bring that fellow around again. After all, all he did was make the bomb. I'm the guy who fired it off."

Truman to Dean Acheson after his meeting with Oppenheimer.[35]

"Some people profess guilt to claim credit for the sin."

The mathematician John von Neumann, following Oppenheimer's well-publicized hand-wringing over his role in developing atomic weapons.[36]

On returning from Bikini [where atomic weapons were tested in July 1946] one is amazed to find the profound change in the public attitude toward the problem of the atomic bomb. Before Bikini the world stood in awe of this new cosmic weapon. . . . Since Bikini this feeling of awe has largely evaporated.

William L. Laurence, a reporter for *The New York Times*, August 1946.[37]

In conjunction with other mass destruction weapons *it is possible to depopulate vast areas of the earth's surface, leaving only vestigial remnants of man's material works.*

General Curtis LeMay in 1947, reporting on the atomic bomb tests on Bikini the year before.[38]

"I'll be damned if I'll let anybody in Washington or any politicians tell me what work not to do."

Norris Bradbury, Oppenheimer's successor at Los Alamos, on being told of pressures in Washington to block work on the "super."[39]

Throughout my recent trip in Europe I was increasingly impressed by the fact that the only balance that we have against the overwhelming manpower of the Russians, and therefore the chief deterrent to war, is the threat of the immediate retaliation with the atomic bomb.

Secretary of Defense James Forrestal, preparing a series of "Points for the President" in November 1948.[40]

"We do not have a secret to give away—the secret will give itself away."

Henry Stimson at a cabinet meeting on September 21, 1945. (It was Stimson's 78th birthday and last day in office as Secretary of War, ending a long career in which he served in the cabinets of four Presidents.)[41]

"In the expansionist circles of the USA a new, peculiar sort of illusion is widespread—faith is placed in the secret of the atomic bomb, although this secret has long ceased to exist."

> Molotov, in a speech on the 30th anniversary of the Russian Revolution, November 1947.[42]

"I was astonished that it [the first Russian atomic test] came that soon. I will tell you this was a peculiar kind of psychology. If you had asked anybody in 1944 or 1945 when would the Russians have it, it would have been five years. But every year that went by you kept on saying five years."

> Physicist I. I. Rabi, describing the shock of American officials on learning of the first Russian bomb test in September 1949.[43]

The general question was "What now?" At once I said that work should be pushed on the "super."

> The mathematician Stanislaw Ulam, describing a meeting at Los Alamos with John von Neumann and Edward Teller after learning of the first Soviet atomic bomb test in September 1949. The "super" was their term for the hydrogen or thermonuclear bomb.[44]

"I hope and pray we will never make the super."

> Oppenheimer, shortly before President Truman's order to begin a crash program in January 1950 to make the super.[45]

What he [Senator Brien McMahon] is talking about is the inevitability of war with the Russians, and what he says adds up to one thing: blow them off the face of the earth, quick, before they do the same to us—and we haven't much time.

> Atomic Energy Commission Chairman David Lilienthal, in his diary, November 1, 1949. McMahon strongly favored, and Lilienthal opposed, an all-out program to develop the thermonuclear or hydrogen bomb. President Truman announced a program to develop the new weapon on January 31, 1950.[46]

Gleb Wataghin: "Hello, Johnny. I suppose you are not interested in mathematics any more. I hear you are now thinking about nothing but bombs."

John von Neumann: "That is quite wrong. I am thinking about something much more important than bombs. I am thinking about computers."

> John von Neumann responding to an old friend in 1946. His success with computers opened the way for complex theoretical calculations which led directly to the invention of thermonuclear weapons.[47]

In spite of an initial, hopeful-looking "flare up," the whole assembly started slowly to cool down. Every few days Johnny [von Neumann] would call in some results. "Icicles are forming," he would say dejectedly.

> Ulam, on early failures of the H-bomb program.[48]

Total War

"There will be dancing in the streets of Los Alamos tonight."

> The mathematician John von Neumann, after an April 1954 test failure of a weapon developed by Edward Teller at his rival Livermore laboratory.[49]

"If I were the Reds, I would fill the oceans all over the world with radioactive fish. It would be so easy to do!"

> Lewis Strauss, chairman of the Atomic Energy Commission, to Eisenhower's press secretary, James Hagerty, following the *Lucky Dragon* incident when a Japanese fishing boat was contaminated with fallout after a thermonuclear test on February 28, 1954. The device yielded an explosive force equivalent to fifteen million tons (or "megatons") of TNT and remains the most powerful weapon ever detonated by the United States. The expected yield was less than half the actual result.[50]

"I'm not sure I believe in hell."

> John Foster Dulles, U.S. Secretary of State 1953–1959, on being told that observing an atomic test is "like having a look at hell." He refused to pass on a recommendation that Eisenhower witness one.[51]

I had to explain to him [Marshal A. A. Grechko] that the smaller the explosive charge of a warhead, the more raw [fissionable] material you need—and we simply didn't have enough raw material to go around.

Nikita Khrushchev explaining why Russia could not build large numbers of tactical nuclear weapons in the late 1950's.[52]

"I remember one day when he [Kennedy] asked me what happens to the radioactive fallout, and I told him it was washed out of the clouds by the rain and would be brought to earth by the rain. And he said, looking out the window, 'You mean, it's in the rain out there?' and I said, 'Yes.' He looked out the window, very sad, and didn't say a word for a few minutes."

Jerome Weisner, President Kennedy's science adviser.[53]

"I made one great mistake in my life—when I signed the letter to President Roosevelt recommending that atom bombs be made, but there was some justification—the danger that the Germans would make them."

Einstein to Linus Pauling, July 1969.[54]

2

THE MISSILES

We must give an account of the metal known as iron, the most useful and the most fatal in the hand of mankind. For by the aid of iron we lay open the ground, we plant trees, we prepare our vineyard trees, and we force our vines each year to resume their youthful state, by cutting away their decayed branches. It is by the aid of iron that we construct houses, cleave rocks, and perform so many other useful offices of life. But it is with iron also that wars, murders and robberies are effected, and this, not only hand to hand, but from a distance even, by the aid of missiles and winged weapons, now launched from engines, now hurled from the human arm, and now furnished with feathery wings. This last I regard as the most criminal artifice that has been devised by the human mind: for, as if to bring death upon man with still greater rapidity, we have given wings to iron and taught it to fly. Let us therefore acquit Nature of a charge that belongs to man himself.

Pliny the Elder (A.D. 23–79), *Natural History.*[1]

Total War

When Archidamus saw a new war engine shooting a missile he said: It is all over with human courage.

> Plutarch (A.D. 46–120).[2]

O! curs'd device! base implement of death! Fram'd in the black Tartarean realms beneath! By Beelzebub's malicious art design'd to ruin all the race of humankind.

> Ludovico Ariosto (1474–1533) on gunpowder.[3]

I am always afraid that they will eventually succeed in discovering some secret which will provide a quicker way of making men die, and exterminate whole countries and nations.

> Montesquieu, *Persian Letters* (1721).[4]

If men were all virtuous, returned the artist, I should with great alacrity teach them all to fly. But what would be the security of the good, if the bad could at pleasure invade them from the sky? Against an army sailing through the clouds neither walls, nor mountains, nor seas, could afford any security.

> Samuel Johnson, *The History of Rasselas, Prince of Abyssinia* (1759).[5]

The Missiles

"[A rocket is] ammunition without ordnance; it is the soul of artillery without the body."

The British inventor William Congreve (1772–1828), creator of the modern war rocket.[6]

"I don't want to set fire to any town, and I don't know any other use of rockets."

The Duke of Wellington, following the burning of Copenhagen by twenty-five thousand British rockets in 1806.[7]

"I don't understand how a serious scientist or engineer can play around with rockets."

Engineer Vannevar Bush remark to Robert Millikan and aerodynamicist Theodor von Kármán in the 1930's.[8]

"Kármán can take the Buck Rogers job."

Remark of MIT scientist Jerome Hunsaker in the fall of 1938. Hunsaker concentrated on something he thought more important than rockets—the problem of de-icing windshields for the Army Air Corps.[9]

"Kármán, do you honestly believe that the air corps should spend as much as $10,000 for such a thing as rockets?"

Major Benjamin Chidlaw, an aide to General Henry A. "Hap" Arnold, to Theodor von Kármán in the fall of 1939, when the Army Air Corps was planning its first contract for rocket research.[10]

"We must have the long-haired professors in von Kármán's board see all the gadgets and data and drawings so as to give us a Buck Rogers program to cover the next twenty years."

General "Hap" Arnold in 1945, speaking of the Air Force Scientific Advisory Group set up by the CalTech aerodynamic engineer, Theodor von Kármán (1881–1963), to advise the Air Force on future development.[11]

"Why didn't I believe in the success of your work? If we had had this weapon in 1939, we never would have had this war. Now and in the future, Europe and the world are too small for a war. With such weapons available war will become unbearable for the human race."

Hitler to General Walter Dornberger, commander of the V-2 rocket program in July 1943.[12]

"There has been a great deal said about a 3,000-mile high-angle rocket.

"In my opinion such a thing is impossible and will be impossible for many years. The people who have been writing these things that annoy me have been talking about a 3,000-mile high-angle rocket shot from one continent to another carrying an atomic bomb, and so directed as to be a precise weapon which would land on a certain target such as this city.

"I say technically I don't think anybody in the world knows how to do such a thing, and I feel confident it will not be done for a very long period of time to come. I think we can leave that out of our thinking."

Vannevar Bush, chief of the Office of Scientific Research and Development, in a December 1945 Senate hearing.[13]

An analysis of the performance of the V-2 proves conclusively that it is possible to build liquid-fuel rockets of almost any size. There is little question that their size, range, and accuracy will be enormously improved. We must look forward to the possibility, in time, of our cities being accurately bombed from across the seas.

Los Alamos scientist Harrison Brown, *Must Destruction Be Our Destiny?* (1946).[14]

The AAF [Army Air Force] *must* go to guided missiles for the initial heavy casualty phases of future wars.

General Curtis LeMay, in a memo to General Carl Spaatz, September 1946.[15]

Total War

Warfare will then take on a Krakatoan aspect.

> Major General J.F.C. Fuller on the effect of the marriage of atomic weapons with missiles in 1946. [16]

"This is absolutely intolerable. We defeated Nazi armies; we occupied Berlin and Peenemünde; but the Americans got the rocket engineers. What could be more revolting and more inexcusable? How and why was this allowed to happen?"

> Stalin to Colonel General I. A. Serov in the late 1940's. [17]

"Do you realize the tremendous strategic importance of machines of this sort? They could be an effective straitjacket for that noisy shopkeeper Harry Truman. We must go ahead with it, comrades. The problem of the creation of transatlantic rockets is of extreme importance to us."

> Stalin at a Politburo meeting in the Kremlin on March 15, 1947, as reported by the Soviet expert G. A. Tokaty, who defected the following year. (Russian, like American, rockets were not in fact "transatlantic," but were aimed to pass over the arctic.) [18]

Whether we can get the 1500 mile or the 5000 mile first, I do not know, but we should try for both. Whoever gets either one first will have the world by the tail with a downpull.

The Missiles

Statesman and financier Bernard Baruch, writing of the U.S. missile program in a letter to Eisenhower, February 6, 1956. In the event, the U.S. got both first. One Russian response was Khrushchev's secret attempt to deploy missiles in Cuba in the fall of 1962. [19]

During dinner Mrs. Eden asked us, "Tell me, what sort of missiles do you have? Will they fly a long way?"

"Yes," I said, "they have a very long range. They could easily reach your island and quite a bit farther."

Khrushchev to the wife of the British prime minister at a dinner party in London, April 1956. [20]

"That's not what he said—he said 'imminent.' "

An American scientist, whispering to William Pickering at a meeting in Washington on September 30, 1957, after a Russian scientist had just said—according to the official translator—that Russia expected to launch a satellite "in the near future." [21]

"We are ahead of the Russians in supercarriers, tomahawks and scalping knives. No one questions our pre-eminence in any of these three categories."

Representative Clarence Cannon to Admiral Arthur A. Radford, Chairman of the Joint Chiefs of Staff, at a Congressional hearing, February 1958. [22]

"Your generals talk of maintaining your position in Berlin with force. That is bluff. If you send in tanks, they will burn and make no mistake about it. If you want war, you can have it, but remember it will be your war. Our rockets will fly automatically."

> Khrushchev to Averell Harriman in June 1959 during a crisis over Berlin. Harriman said Khrushchev's "colleagues around the table chorused the word 'automatically.' "[23]

"Put your hand on the back of your neck. All right. You feel it? That's your neck. Well, that's what we're trying to save. That's what this program is all about."

> Pep talk by Rear Admiral William Raborn, director of the Navy's Polaris program in the 1950's.[24]

"Today Charleston becomes a potent thorn in the side of the Soviet Union. Today Charleston becomes the deterrent capital of the world. We make an indelible mark on the planning maps of the Soviet Union. This is the penalty we must pay for this great honor."

> Congressman L. Mendel Rivers in a speech at the dedication of the Polaris base in Charleston, South Carolina, March 29, 1960.[25]

I found myself in the difficult position of having to debate on a course of action which would answer the American threat but which would also avoid war. Any fool can start a war, and once he's done so, even the wisest of men are helpless to stop it—especially if it's a nuclear war.

Khrushchev, recalling his decision to install Soviet missiles in Cuba in 1962.[26]

"I will always remember Dean Acheson coming into our meeting and saying that he felt that we should knock out Soviet missiles in Cuba by an air strike. Someone asked him, 'If we do that, what do you think the Soviet Union will do?' He said, 'I think I know the Soviet Union well. I know what they are required to do in the light of their history and their posture around the world. I think they will knock out our missiles in Turkey.' And then the question came again. 'Well, what do we do?' 'Well,' he said, 'I believe under our NATO treaty with which I was associated, we would be required to respond by knocking out a missile based inside the Soviet Union.' 'Well, then what do they do?' 'Well,' he said, 'then that's when we hope cooler heads will prevail and they'll stop and talk.' "

Theodore Sorensen, an aide to John F. Kennedy, describing a meeting of the Executive Committee during the Cuban Missile Crisis in 1962. Cooler heads prevailed in Washington, and Kennedy chose to blockade Cuba rather than launch an air strike.[27]

"There was a smell of burning in the air."

Khrushchev, after the Cuban Missile Crisis in October 1962.[28]

"You Americans will never be able to do this to us again."

> The Soviet official Vasily V. Kuznetzov to John J. McCloy following the Cuban Missile Crisis in October 1962. Kuznetzov was negotiating the withdrawal of Soviet missiles.[29]

"Don't kid yourself."

> President Richard Nixon to an interviewer who asked in 1985 if he agreed with officials who said the Soviet decision to withdraw missiles from Cuba in October 1962 had nothing to do with American nuclear superiority.[30]

There is indeed grave risk in using ballistic missiles, but that risk is not uncertainty of accuracy.

> Charles Stark Draper, the Einstein of missile guidance, *The New York Times*, September 20, 1981.[31]

"We made the mistake under the previous leadership of relying on nuclear weapons. This meant we were left far behind in the development of our traditional capabilities. So now we have no fighter-bomber that can match the Americans. We shall catch them up, but at present there's a gap."

44

The Missiles

Soviet leader Leonid Brezhnev, responding to an Egyptian request for aircraft to match the Israeli Phantoms.[32]

"I wish I had never heard of this weapon."

President Jimmy Carter in a White House meeting to discuss basing modes for the new MX missile, March 20, 1978.[33]

3

HOW MUCH IS ENOUGH?

"We want eight and we won't wait."

Slogan of the Conservative Party in Britain in 1908 during a campaign to build more Dreadnought class battleships for the Navy.[1]

In the end a curious and characteristic solution was reached. The Admiralty had demanded six ships: the economists offered four: and we finally compromised on eight.

Churchill on the Dreadnought controversy.[2]

"No nation ever had an army large enough to guarantee it against attack in time of peace, or ensure it victory in time of war."

President Calvin Coolidge.[3]

Total War

In the future, war will be waged essentially against the unarmed populations of the cities and great industrial centres.

The Italian General Giulio Douhet in his influential book *The Command of the Air* (1921).[4]

What Butt was particularly anxious to know was the ratio of casualties per ton of bombs in areas with different housing densities.

Solly Zuckerman, a British expert on bombing, describing a request in 1942 for figures from D. M. Butt, an aide to Lord Cherwell, Churchill's science advisor during World War II.[5]

Sixty percent of the bombs dropped are not accounted for, less than one percent have hit the aiming point and about three percent [land] within 500 feet.

Letter from then Colonel Curtis LeMay to an old friend, January 12, 1943, describing difficulties in bombing German targets accurately.[6]

We should never allow the history of this war to convict us of throwing the strategic bomber at the man in the street.

General Ira C. Eaker, commander of the U.S. 8th Air Force in Britain during World War II, in a letter of January 1, 1945. The Americans eventually followed the British example, however, and commenced bombing raids on residential districts of German cities.[7]

Here is the antidote for qualms about strategic bombing.

Air Force historian Bruce Hopper, on a visit to Buchenwald, April 1945.[8]

"If we were to atomic bomb the Soviet Union, what targets would you choose? Would you bomb Leningrad, with the Hermitage? . . . But if you're really serious about this why is there any question? . . . Go home and think about it."

Secretary of State George Marshall to his aide, Gordon Arneson, who suggested using the bomb to break the Berlin blockade imposed by the Russians in June 1948. At that time the United States still had a monopoly on atomic weapons.[9]

"If civilians are going to be killed, I would rather have them [be] their civilians than our civilians."

Secretary of the Air Force Stuart Symington in October 1949, during an Air Force–Navy argument over the morality of strategic bombing. The Navy insisted it was immoral, and the country should therefore buy aircraft carriers.[10]

"Why should we have a Navy at all? There are no enemies for it to fight except apparently the Army Air Force."

Carl Spaatz, Commanding General of the Army Air Forces, answering his own question, early 1946.[11]

"We consider that strategic air warfare, as practiced in the past and as proposed for the future, is militarily unsound and of limited effect, is morally wrong, and is decidedly harmful to the stability of a post-war world."

Rear Admiral Ralph A. Ofstie to the House Armed Services Committee on October 11, 1949, opposing Air Force plans to build the B-36.[12]

"It's a fundamental law of defense that you always have to use the most powerful weapons you can produce."

Major General James Burns, a special advisor to Secretary of Defense Louis Johnson, after another Air Force officer said in October 1949 that we already had weapons big enough to obliterate Moscow. At that time Johnson favored an all-out program to develop hydrogen bombs, hundreds of times more powerful than atomic bombs already in the U.S. arsenal.[13]

"Gentlemen, you don't have a war plan. You have a war spasm."

The Rand analyst Bernard Brodie, responding to a briefing by officers of the Strategic Air Command on plans for a nuclear war in 1950.[14]

"Local defense will always be important. But there is no local defense which alone will contain the mighty land power of the Communist world. Local defense must be reinforced by the further deterrent of massive retaliatory power."

> Speech by Secretary of State John Foster Dulles at the Council on Foreign Relations, January 12, 1954, introducing the policy of "massive retaliation."[15]

"Well, we did not build those bombers to carry crushed rose petals!"

> General Thomas S. Power, answering a reporter who asked if American B-47's in England carried nuclear weapons—at a NATO meeting in Paris a few weeks after the Soviet Sputnik success on October 4, 1957.[16]

A high-ranking official of the State Department, who was not a career Foreign Service officer, could not conceal his alarm. "We must stop this before we are all burned to a crisp!" he exclaimed. . . . His reaction based on fear was not appreciated by his colleagues.

> Diplomat Robert Murphy, recalling a staff meeting during the Suez crisis in 1956, following Russian missile threats.[17]

"The principle of neutrality . . . has increasingly become an obsolete conception, and, except under very special circumstances, it is an immoral and shortsighted conception."

> John Foster Dulles, in a speech on June 9, 1956.[18]

Total War

"Our real mission, you might say, is to have that Russian planner get up from his table every morning and turn to Mr. Khrushchev and shake his head and say, 'Today is not the day, Comrade.' "

General Thomas S. Power, Commander in Chief of the Strategic Air Command (1957–1964).[19]

"Those people, God damn it, don't know what the progress of technology means. They're still thinking in World War I terms."

President Dwight D. Eisenhower, November 1959, on military objections to reduction of conventional forces.[20]

"Our technology is what will save us."

Secretary of Defense Harold Brown, 1979.[21]

"That is a secret."

Khrushchev to the wife of the U.S. Ambassador to Russia, Llewellyn Thompson, at a New Year's Eve party in the Kremlin on January 1, 1960. After Khrushchev said the USSR had reserved fifty nuclear bombs for Britain and thirty each for France and West Germany, Jane Thompson had asked him, "How many do you have for us?"[22]

"Why don't we go completely crazy and plan on a force of ten thousand?"

> Exasperated remark by President Eisenhower, on being told in April 1960 that the Air Force was still pushing for more Minuteman missiles.[23]

"Damn it, if you keep talking about ten thousand missiles, you are talking about pre-emptive attack. Why don't you just say so?"

> Secretary of Defense Robert McNamara in 1962 to an Air Force general who wanted ten thousand Minuteman missiles.[24]

"General, we'll never have ten thousand ICBMs in place."

> President John F. Kennedy contradicting General Thomas Power of the Strategic Air Command, after the general referred to a future moment *when* the U.S. would have that many missiles. Kennedy was right—so far. But the United States now has roughly ten thousand strategic warheads.[25]

"You cannot make decisions simply by asking yourself whether something might be nice to have. You have to make a judgment on how much is enough."

Secretary of Defense Robert McNamara at a meeting of the American Society of Newspaper Editors in April 1963.[26]

"I don't believe that any time in our lifetime they will reach parity with us in the total power of their system versus ours."

McNamara in a congressional hearing in 1963.[27]

So grim a subject does not exclude an appropriate kind of humor used very sparingly, but levity is never legitimate.

Bernard Brodie in a letter to Herman Kahn in 1962, scolding him for coining the word "wargasm" to describe an all-out nuclear war in his book *On Thermonuclear War* (1960).[28]

Most of the theorizing about the Soviet-American strategic relationship has come from Americans. . . . Soviet views were unknown. The Soviets seldom take up their strategic pens to rationalize the irrational.

Strategic Arms·Limitation Talks (SALT) negotiator Gerard Smith in his memoirs.[29]

"Welcome to the world of strategic analysis, where we program weapons that don't work against threats that don't exist."

How Much Is Enough?

Ivan Selin, chief of the strategic forces division of the Pentagon's Office of Systems Analysis in the 1960's.[30]

"General, I don't think you understand. I didn't come here for a briefing. I came to tell you what we have decided."

Alain C. Enthoven, head of the Systems Analysis office in McNamara's Pentagon, to an Air Force general in Germany in the early 1960's.[31]

"I am profoundly apprehensive of the pipe-smoking, tree-full-of-owls type of so-called professional defense intellectuals who have been brought into this nation's capital."

General Thomas D. White, former Air Force Chief of Staff, in the early 1960's.[32]

A strategic planner must be conservative in his calculations; that is, he must prepare for the worst plausible case and not be content to hope and prepare for the most probable.

Secretary of Defense Robert McNamara, 1967.[33]

In the real world of real political leaders—whether here or in the Soviet Union—a decision that would bring even one

55

hydrogen bomb on one city of one's own country would be recognized in advance as a catastrophic blunder; ten bombs on ten cities would be a disaster beyond history; and a hundred bombs on a hundred cities are unthinkable.

Former National Security Adviser McGeorge Bundy, 1969.[34]

When we came to office in 1969, the estimate of casualties in case of a Soviet *second* strike stood at over fifty million dead from *immediate* effects. . . .

Former Secretary of State Henry Kissinger. A Soviet second strike would of course imply a U.S. first strike.[35]

"I asked them what their smallest target was, and it was ridiculous. Essentially, it was a pasture that returning bombers might use."

The Harvard chemist Paul Doty, who worked for the Joint Strategic Target Planning Staff in the mid-1970's.[36]

"Hell, what are we going to do with all of those?"

Admiral Gerald D. Miller, deputy director of the Joint Strategic Target Planning Staff in the early 1970's, reacting to the sudden rise in available warheads following the MIRVing of Poseidon missiles.[37]

"What we are doing now is targeting a war-recovery capability."

> General George Brown, Chairman of the Joint Chiefs of Staff, 1977. War planners refer to factories, power plants, and transportation and communications facilities as "recovery targets." Attacking them amounts to attacking cities, where most recovery targets are located.[38]

In the nuclear age a bluff taken seriously is useful; a serious threat taken as a bluff may prove disastrous. The longer deterrence succeeds, the more difficult it is to demonstrate what made it work. Was peace maintained by the risk of war, or because the adversary never intended aggression in the first place?

> Henry Kissinger, *White House Years* (1979).[39]

"What in the name of God is strategic superiority? What is the significance of it, politically, militarily, operationally, at these levels of numbers? What do you do with it?"

> Henry Kissinger at a press conference in Moscow, July 29, 1974.[40]

Nuclear superiority was very useful to us when we had it.

> Richard Nixon, *The Real War* (1980).[41]

4

THE RUSSIANS

"Though our country is in no danger of invasion, no sooner is peace concluded than plans are laid for a new war, which has generally no other foundation than the ambition of the Sovereign. . . . In a word, we are so exhausted and ruined by the keeping up of a standing army . . . that the most cruel enemy, though he should devastate the whole Empire, could not cause us one half of the injury."

Complaint of Russian nobles in the time of Peter the Great (1682–1725), as recorded by the Prussian diplomat Vockerodt.[1]

"Everything that may plunge Russia into chaos and make her return to obscurity is favourable to our interests."

Louis XV of France in 1763.[2]

"Sire, there are many roads to Moscow. Charles XII chose that by Poltava."

Total War

Russian ambassador to Napoleon in 1811, after he had been asked for the best road to Moscow. Poltava was the site of Charles XII's crushing defeat by the Russians in 1709.[3]

"I have come to finish off, once and for all, the colossus of the North. . . . They must be thrust back into their snow and ice so that, for a quarter of a century at least, they will not be able to interfere with civilized Europe."

Napoleon, at the outset of his invasion of Russia in 1812.[4]

We are one of those nations which do not appear to be an integral part of the human race, but exist only in order to teach some great lesson to the world. Surely the lesson we are destined to teach will not be wasted; but who knows when we shall rejoin the rest of mankind, and how much misery we must suffer before accomplishing our destiny?

Pyotr Chaadayev of Russia, in 1829.[5]

There are now two great nations in the world which, starting from different points, seem to be advancing toward the same goal: the Russians and the Anglo-Americans. . . .

All other peoples seem to have nearly reached their natural limits. . . . They alone march easily and quickly forward along a path whose end no eye can yet see.

The American fights against natural obstacles; the Russian is at grips with men. The former combats the wilderness and barbarism; the latter, civilization with all its arms. America's conquests are made with the plowshare, Russia's with the sword.

The former relies on personal interest and gives free scope to the unguided strength and common sense of individuals.

The latter in a sense concentrates the whole power of society in one man.

One has freedom as the principal means of action; the other has servitude.

Their point of departure is different and their paths diverse; nevertheless, each seems called by some secret design of Providence one day to hold in its hands the destinies of half the world.

Alexis de Tocqueville, *Democracy in America* (Volume I, 1835).[6]

Russia sees Europe as a prey which our dissensions will sooner or later deliver up to her. . . . It is the history of Poland recommencing on a larger scale.

The Marquis de Custine in his account of a trip through Russia in 1839.[7]

As long as Russia does not exceed her natural limits, the Russian navy will be the plaything of emperors—nothing more! . . . [It] seemed to me just a parade. But heaven and the Russians know what a pleasure a parade is! The taste for reviews is pushed in Russia to the point of madness.

> The Marquis de Custine, on observing Imperial Navy maneuvers in July, 1839.[8]

The Russian government is the discipline of the camp substituted for the civic order—it is a state of siege become the normal state of society.

> Marquis de Custine, *Journey for Our Time*—his diary of a trip taken in 1839.[9]

It may be doubted whether a cabinet council often takes place now in our Foreign Office without Russia being uppermost in every English statesman's thoughts.

> The British historian Edward Shepherd Creasy, reflecting on the Battle of Pultowa (1709) in his book, *The Fifteen Decisive Battles of the World* (1851), first published shortly before the Crimean War.[10]

The idea of Russian diplomatic supremacy owes its efficacy to the imbecility and timidity of the Western Nations, and . . . the belief in Russia's superior military power is hardly less a delusion.

> Karl Marx in the *New York Tribune,* December 30, 1853.[11]

Scientific socialism, the socialism of the West, bounces off the Russian masses like a pea off the wall.

> Sergey Kravchinsky of the *narodniki,* a revolutionary group in the 1870's.[12]

The state swelled up; the people grew lean.

> The Russian historian V. O. Kliuchevsky, describing the growth of the czarist state in 1911.[13]

"The future belongs to Russia. She keeps growing and growing, bearing down on us like a nightmare."

> Theobald von Bethmann-Hollweg, Chancellor of Germany, on July 7, 1914.[14]

"The permanent coexistence of the Soviet Republic and the imperialist states is unthinkable. One or the other must triumph in the end. And before that end arrives, a series of terrible collisions between the Soviet Republic and the bourgeois nations is inevitable."

> Lenin.[15]

I maintain that we received strategic warning of the Communists' intentions as early as 1848—over one hundred years ago—when Karl Marx and Friedrich Engels published their *Communist Manifesto.*

General Thomas S. Power, Commander in Chief of the Strategic Air Command, 1957–1964.[16]

"To choose one's victim, to prepare one's plan minutely, to slake an implacable vengeance, and then to go to bed. . . . There is nothing sweeter in the world."

Stalin to Kamenev and Feliks Dzerzhinski in 1923.[17]

But Russians, in their very nature, it seems, pass quickly from the highest enthusiasm to complete dejection and distrust. . . . Incurable pessimists, they actually dislike to look forward with hope, and prefer to expect of the future nothing but misfortune.

Marie, Grand Duchess of Russia, *Education of a Princess* (1930).[18]

"*Avant la révolution, les Russes étaient très tristes. Maintenant ils sont tristes.*"
　　　["Before the Revolution, Russians were very sad. Now they are sad."]

Soviet acquaintance to Edmund Wilson during his trip to Russia in 1935.[19]

The Russians

He [General John Dill, Director of Military Operations in Britain] clearly disliked the idea that we might be on the side of Russia. . . . Could we not let Germany expand eastwards at Russia's expense? I suggested that while it might pay from a short view, it was dubious from a long view—there was a danger in feeding the tiger that might turn on you.

> B. H. Liddell Hart, in his notes on their discussion of defense problems in the spring of 1935.[20]

"I cannot forecast to you the action of Russia. It is a riddle wrapped in a mystery inside an enigma."

> Winston Churchill radio talk, October 1, 1939.[21]

"Did you punch him in the snout? If he does something like that again, punch him right in the snout!"

> Stalin to a general who had been complaining of a subordinate.[22]

"Zuck, when we've finished with Germany, we'll still want you when we take on the Russians. Don't forget."

> An American general in Sicily to the British bombing expert, Solly Zuckerman, in December 1943.[23]

"What are we going to have between the white snows of Russia and the white cliffs of Dover?"

> Winston Churchill in February 1945, wondering gloomily about Britain's strategic position after Allied bombers had completed the destruction of Germany.[24]

"We must keep the Russians out of Denmark."

> General Sir Miles Dempsey to General James Gavin in April 1945.[25]

Mr. Byrnes's . . . view [was] that our possessing and demonstrating the bomb would make Russia more manageable in Europe.

> Physicist Leo Szilard, after a meeting with James Byrnes in the spring of 1945.[26]

Oppenheimer: "Don't you think if we tell the Russians what we intend to do and then use the bomb in Japan, the Russians will understand it?"
Szilard: "They'll understand it only too well."

> Spring 1945.[27]

"We have decided to go our own way."

> Stalin to Averell Harriman in October 1945.[28]

"What would have happened if we hadn't made it then? They would have shot [us]. Just shot."

> Vassily Emelyanov, a Russian physicist who worked on the first Soviet atomic bomb.[29]

"These Americans represent the new Roman Empire and we Britons, like the Greeks of old, must teach them how to make it go."

> Harold Macmillan, Churchill's personal representative in Algiers in 1943, to his staff.[30]

They sometimes have a tendency to be caught in their own propaganda which, as you know, is to the effect that imperialist America is rushing around to take over the world.

> Secretary of State George C. Marshall, 1948.[31]

"The question is asked: 'What will happen when they [the Russians] get the atomic bomb themselves and have accumulated a large store?'

"You can judge for yourself what will happen then by what is happening now. If these things are done in the green wood what will be done in the dry?"

> Winston Churchill in October 1948.[32]

"The country needs aircraft like black bread. You can offer pralines, cakes and so on, but there's no point—there aren't the ingredients to make them out of."

> The Soviet aircraft designer Andrey Tupolev. The first Soviet long-range bombers were carbon copies of American B-29's seized when they landed in Siberia after a raid on Japan in 1945.[33]

We wished every success to Kim Il-sung and toasted the whole North Korean leadership, looking forward to the day when their struggle would be won.

> Khrushchev, describing a dinner in Moscow before the invasion of South Korea in June 1950.[34]

He said Russian peasants knew how to deal with them—they killed them.

> K.P.S. Menon, the Indian Ambassador to the Soviet Union, reporting Stalin's remark when Menon discovered him doodling wolves with a red pencil, February 1953.[35]

Dulles immediately led me to his study to talk. He was excited about the prospects of Beria's arrest's setting off a bloody struggle for power that might lead to the overthrow of the Soviet regime.

> Charles Bohlen, the American ambassador to the Soviet Union, on being summoned to Washington in July 1953 by Secretary of State John Foster Dulles.[36]

"There is only one real question that interests these people, I mean the Soviet leaders, and that is the question of who has the ability to haul people out of bed at three in the morning and cause them to disappear without giving any accounting for them."

> George Kennan, former ambassador to the Soviet Union, April 20, 1954.[37]

"We are stuck fast in secrecy like a fly in treacle."

> Admiral A. I. Berg, Soviet Deputy Minister of Defense for Radar and Radio Engineering (1953–1957).[38]

"Don't tease the geese."

> Russian proverb cited by Khrushchev in describing a border dispute with China in 1954.[39]

Total War

At night they show distinguished foreign guests in the Kremlin, on a strictly confidential basis, films of Soviet hydrogen bombs exploding, and the next day they conduct negotiations with the same guests about mutual cooperation and the preservation of peace.

> Veljko Mićunović, Yugoslavia's ambassador to the Soviet Union, who noted that this "diplomatic device" worked on Tito and the Shah of Iran, July 1956.[40]

"We now have warheads which have so powerful a yield that we couldn't deliver them on Germany, because the fallout would contaminate the Soviet Union. When President Nasser was our guest we showed him a film of an atomic explosion. But that was in 1958. We were proud of it then, but now I feel ashamed. It was just a tiny thing, the size of a pea compared with the giants we've got now."

> Khrushchev, speaking to the Egyptian journalist Mohammed Heikal in May 1964. "All this was said with a typical peasant's enthusiasm," commented Heikal.[41]

"We will bury you."

> Khrushchev, speaking at the Polish Embassy in Moscow, November 18, 1956. Khrushchev's remark was widely taken as a threat; he himself apparently meant only to say that socialism would attend the funeral of capitalism.[42]

The Russians

"The Soviet Union will continue to fight stubbornly for peace, which we have especial need of for the next fifteen or twenty years. After that no one will be able to go to war even if he wants to."

> Khrushchev, in a farewell visit with the Yugoslavian ambassador Veljko Mićunović, October 1958.[43]

We exchanged opinions in the leadership and decided to go ahead as a concession to the military, which was in favor of these ships. Our naval commanders thought they looked beautiful and liked to show them off to foreigners. An officer likes to hear all the young sailors greet his command with a loud cheer.

> Khrushchev, on the decision to build four modern cruisers in the late 1950's.[44]

"Ah yes. And are your officers also in the Soviet ministries?"

> Raúl Castro, Fidel's brother and leader of a Cuban delegation to Czechoslovakia in the fall of 1959, to General Bohimir Lomsky, the Czech Minister of Defense, who had just admitted that Soviet officers were assigned to watch things in the many offices of the Czech Ministries of Defense and the Interior.[45]

"Russians."

> The physicist Edward Teller's answer, when asked what he thought we would find on the moon.[46]

Total War

The tragedy of contemporary Russia is that the whole elite of the nation, its intelligentsia, its civil services, and all its politically minded elements share in one degree or another in Stalin's guilt.

Isaac Deutscher, *Ironies of History* (1966).[47]

"The Soviet Union will decide whether or not Austria can remain neutral. Not the Austrian Government."

General Alexey A. Yepishev, chief political commissar of the Soviet armed forces, to Jan Sejna (later a defector) in Moscow, March 1967.[48]

"I mean *Russian* Russians."

Zbigniew Brzezinski, President Carter's national security advisor, clarifying his question about casualties during a 1977 discussion of nuclear targeting, after a briefing officer told him how many Soviets—not how many ethnic Russians—would be killed in a U.S. nuclear strike.[49]

Q:"What do you do in the case of nuclear attack?"
A:"Wrap yourself in a sheet and crawl slowly to the cemetery."
Q:"Why slowly?"
A:"So as to avoid causing a panic."

Muscovite joke based on the acronym derived from the Russian words for civil defense, GROB, which is also their word for coffin.[50]

The Russians

On what does the effectiveness of creativity on the battlefield depend? It depends on a firm knowledge of the regulations.

Red Star, the official newspaper of the Soviet armed forces, 1971.[51]

Communist policy is often described as diabolically clever, complicated, following well-thought-out routes toward world domination. This was not my impression. On the contrary, I found Soviet diplomacy generally rigid; nor is subtlety the quality for which Soviet diplomacy will go down in history. . . . But Soviet diplomacy has one great asset. It is extraordinarily persevering; it substitutes persistence for imagination.

Henry Kissinger, *White House Years* (1979).[52]

"[The Soviets combine] the charm and lightness of the Germans, the openness of the Albanians, the humility of the Indians, and the efficiency of the Latins."

An unnamed foreign leader's comment reported by Seweryn Bialer in *Stalin's Successors* (1980).[53]

"Somebody said the second most stupid thing in the world that a man could say was that he could understand the Russians. I've often wondered what in hell was the first."

President Ronald Reagan during an interview before his meeting in Iceland with General Secretary Gorbachev in October 1986.[54]

5

ARMS CONTROL

There is no secret, there is no defense, and there is no possibility of control except through the aroused understanding and insistence of the peoples of the world.

Einstein.[1]

Diplomacy without armaments is like music without instruments.

Frederick the Great of Prussia (1712–1786).[2]

Threat systems are the basis of politics as exchange systems are the basis of economics.

The American economist Kenneth Boulding, 1968.[3]

Total War

I saw prevailing throughout the Christian world a license in making war of which even barbarous nations would have been ashamed; recourse was had to arms for slight reasons, or for no reason; and when arms were once taken up, all reverence for divine and human law was thrown away; just as if men were thenceforth authorized to commit all crimes without restraint.

Hugo Grotius, *De Jure Belli ac Pacis* (1625).[4]

The sacred truths of Christianity were set forth by Christ and his Apostles some two thousand years ago and the vast majority of people are utterly indifferent to them. . . . In the same manner, it will require many centuries before the idea of the peaceful settlement of conflicts between nations will have any practical consequences.

Count Sergei Witte, advising Nicholas II of Russia not to expect too much at the first Hague Conference in 1899.[5]

Was it a Peace Conference or a War Conference that took place in 1907?

Marschall von Bieberstein, at the conclusion of a stormy conference on disarmament at The Hague in October 1907.[6]

 "Technical arguments are only political arguments dressed up in uniform."

Diplomat Salvador de Madariaga at the Geneva disarmament talks in the 1920's.[7]

He regarded disarmament not so much as a means of preventing war as a way of making it both less damaging to humanity and better for the military artist.

> B. H. Liddell Hart, commenting on General Werner von Blomberg, chief military adviser to the German delegation at the 1932 Disarmament Conference in Geneva. Liddell Hart found him "exceptionally genuine, if boyishly naive," and noted that his fellow officers called him the "Rubber Lion."[8]

I am convinced that extreme simplicity is essential. The trouble is that the moment an apparently simple scheme is produced, the logical Latin mind gets to work. Committees are appointed, every difficulty and possible evasion is carefully scrutinised. In the end it is invariably swamped by a mass of verbiage, definitions, exceptions, technicalities, and theoretical objections, under which the original proposal has completely disappeared.

> Major General A. C. Temperley, British delegate to the Geneva disarmament talks in the 1930's.[9]

So far as the chief causes of the failure of the Conference are concerned, they can be put in a sentence. It was the impossibility of reconciling French demands for security with the German demand for equality of rights.

> Major General A. C. Temperley, on the collapse of the Geneva disarmament conferences in 1934.[10]

Total War

How I enjoyed translating Truman's sentences!

Charles Bohlen, about President Truman's tough talk with Molotov on the Soviets' failure to live up to the Yalta agreements, April 1945.[11]

"I want to express the utmost sympathy with the people who have to grapple with this problem and in the strongest terms to urge you not to underestimate its difficulty."

Oppenheimer, describing the elusiveness of a workable scheme for the international control of atomic weapons, in a farewell speech to scientists at Los Alamos, November 1945.[12]

A situation dangerous to our security could result from impressing on our own democratic peoples the horrors of future wars of mass destruction while the populations of the police states remain unaware of the terrible implications.

Joint Chiefs of Staff to Baruch on June 5, 1946.[13]

I like Mr. Baruch's idea of calling the bomb our "winning weapon"—a weapon which we give up only when we are sure the world will remain safe. . . . If we cannot be sure, we must arm to the teeth with the winning weapon.

Major General Thomas Farrell in August 1946. Farrell had been an aide to General Leslie Groves in the Manhattan Project which developed the first bomb.[14]

"I concluded that I would drop the scientists because as I told them, I knew all I wanted to know. It went boom and it killed millions of people and I thought it was an ethical and political problem and I would proceed on that theory."

Bernard Baruch to Vannevar Bush in March 1946. Baruch presented the first American proposal for international control of the bomb to the UN in June 1946. [15]

"That was the day I gave up hope, but that was not the day for me to say so publicly."

Oppenheimer, about the appointment of Baruch on March 16, 1946, to present the American plan for the international control of atomic weapons to the UN. Oppenheimer, who had helped to write the plan, declined to serve Baruch as an adviser. [16]

"We are here to make a choice between the quick and the dead. . . . Let us not deceive ourselves: we must elect world peace or world destruction."

Bernard Baruch, presenting a plan to the UN for international control of the bomb, June 14, 1946. The Baruch plan was rejected by the Russians. [17]

Bohr called on Secretary of State Dean Acheson to discuss with him the content of his planned [open] letter [to the United

Nations]. The meeting began at, say, two o'clock, Bohr doing all the talking. At about two-thirty Acheson spoke to Bohr about as follows. "Professor Bohr, there are three things I must tell you at this time. First, whether I like it or not, I shall have to leave you at three for my next appointment. Secondly, I am deeply interested in your ideas. Thirdly, up till now I have not understood one word you have said." Whereupon, the story goes, Bohr got so enraged that he waxed eloquent for the remainder of the appointment.

> The physicist Abraham Pais, in *A Tribute to Niels Bohr on the Hundredth Anniversary of his Birth* (1985). Those who knew Bohr may doubt he ever "waxed eloquent" in such circumstances. The story otherwise rings true.[18]

"I've listened to Oppie as carefully as I know how, but I don't know what the hell he's talking about. How do you disarm an adversary by example?"

> Secretary of State Dean Acheson to an aide, after a meeting with Oppenheimer in October 1949.[19]

"I am of the opinion we'll never obtain international control. Since we can't obtain international control we must be strongest in atomic weapons."

> Harry Truman to his advisers during a discussion of arms control in 1949.[20]

Arms Control

We may be likened to two scorpions in a bottle, each capable of killing the other, but only at the risk of his own life.

J. Robert Oppenheimer, 1953.[21]

The Superpowers often behave like two heavily armed blind men feeling their way around a room, each believing himself in mortal peril from the other whom he assumes to have perfect vision.

Henry Kissinger, *White House Years* (1979).[22]

There remains a last argument. Atomic equality may soon result in the neutralization of the terrifying weapon. When it does, the superiority of strategic position, of land forces, and of human potential will come into full play in favor of the Soviet Union.

Then the permanent armies of the West will not merely have to threaten reprisals in order to stifle any fancy for aggression, but will have to be capable of stopping the Russian armies on the two critical battlefields of Western Europe and the Middle East.

Raymond Aron, *The Century of Total War* (1954).[23]

I noted that the time during which a start on missile controls was feasible was rapidly running out and that a few years hence, when comparatively compact solid-propellant ICBMs became available, they will be as difficult to control as nuclear warheads.

> George Kistiakowsky, Eisenhower's science adviser, in his diary for December 1, 1959, describing a meeting of the National Security Council on disarmament.[24]

"When people get bored with it."

> The nuclear physicist I. I. Rabi, asked by Gerard Smith in the late 1950's when the arms race would end.[25]

"Have the U.S. Government reached the point reached by the British, French and Germans in 1914? Do they now believe that only armaments can make them safe, and that keeping the lead in weapons and in force is the only way to safeguard the national interest and uphold the peace?"

> The British disarmament advocate Phillip Noel-Baker in 1955, reacting to the Eisenhower administration's repudiation of previous U.S. disarmament positions.[26]

What you have to remember is that when I faced the problem of disarmament, we lagged significantly behind the US in both warheads and missiles, and the US was out of range for our bombers. . . . That's why I was convinced that as long as the US held a big advantage over us, we couldn't submit to international disarmament controls.

Khrushchev in his memoirs.[27]

"We can stay ahead only by moving ahead."

Edward Teller, explaining opposition to a test ban in 1958.[28]

"It would be very difficult to satisfy Dr. Teller in this field."

President Kennedy, commenting on Teller's opposition to the Test Ban Treaty of 1963.[29]

"I find myself profoundly in anguish over the fact that no ethical discussion of any weight or nobility has been addressed to the problem of atomic weapons. . . . What are we to make of a civilization which has always regarded ethics an essential part of human life . . . [but] which has not been able to talk about the prospect of killing almost everybody, except in prudential and game-theoretic terms?"

J. Robert Oppenheimer, 1960.[30]

"I hope I never hear of a missile gap again."

> Robert McNamara, to his predecessor as Secretary of Defense, Thomas Gates, after a few weeks in office, 1961. The alleged "missile gap" had been a major issue in the 1960 campaign, but once in office President Kennedy and McNamara learned that the United States was, in fact, a long way ahead of the Russians. The latter did not close the gap until the mid-1970's.[31]

"The trouble is not that the Soviets and Americans do not have the same positions; the trouble is that they do not have them at the same time."

> Lawrence D. Weiler of the Arms Control and Disarmament Agency.[32]

Even today there are those in places of influence who have convinced themselves that a cooperative understanding with Moscow and even with Peking can be arranged.

> Robert Murphy, *Diplomat Among Warriors* (1964). Murphy had little faith in such agreements.[33]

"The United States must realize that in both physics and politics each action causes a corresponding counteraction."

> Soviet Premier Alexei Kosygin.[34]

"If we have an arms control agreement, the Russians will cheat. If we have an arms race, we will win."

> General Earle Wheeler, chairman of the Joint Chiefs of Staff, in discussion with a journalist about the first test of MIRVs—Multiple Independently targetable Re-entry Vehicles—on August 16, 1968.[35]

"[The Soviets] are going for a first strike capability, there is no question about that."

> Secretary of Defense Melvin Laird in a closed Congressional hearing, March 22, 1969.[36]

If either side is striving for or appears to be striving for an effective counterforce first strike capability, then there is no hope for strategic arms control.

> SALT negotiator Gerard Smith in a memo to Secretary of State William Rogers, May 1969.[37]

Qualitative limitations are much more difficult to negotiate than quantitative, since it seems to be an article of faith in military thinking that the advance of technology cannot and should not be stopped.

> Gerard Smith on the SALT negotiations in the summer of 1971.[38]

"If it is the view of the Joint Chiefs of Staff that there is no threat to the survivability of Minuteman, it is going to be hard for you to argue that the other elements in the Triad have to be upgraded to compensate for a threat to Minuteman. But, if your position is that Minuteman is threatened, it's going to be hard for you to defend a SALT agreement that scraps the ABM defense of Minuteman. So which is it?"

> Senator Henry Jackson in a Senate Armed Services committee hearing, July 1972. The chiefs backed the Anti-Ballistic Missile treaty, and then pressed for "modernization" of the Triad.[39]

"I don't want my epitaph to say I defended the neutron bomb."

> Adrian Fisher of the Arms Control and Disarmament Agency (ACDA), after defending the neutron bomb at Geneva in 1978.[40]

"SALT fits into our broader effort to enhance national security, an effort which we pursue not only through improving our own forces, but also, where appropriate, through arms control. . . . In this context, I can assure you that we will never constrain our ability to meet our national security needs. A satisfactory SALT agreement will allow us to maintain the effectiveness of the United States strategic arsenal as a deterrent against nuclear war, based on a credible retaliatory capability in the event that war should break out."

> Zbigniew Brzezinski, speaking at a meeting of the Foreign Policy Association, December 20, 1978.[41]

Arms Control

During my years with the Atomic Energy Commission and the State Department I tried to find some moral basis for the use of and even for planning to use nuclear weapons. I could not reconcile my understanding of Christian belief about justifiable use of force and my knowledge of the blast, heat and radiation effects of nuclear explosions. I finally gave up the effort, suspended judgment and kept my conscience quiet with the conclusion that there was no alternative. . . . Recently I was asked to lecture on the moral use of nuclear weapons. I declined. I knew of none.

Gerard Smith, in *Doubletalk: The Story of the First Strategic Arms Limitation Talks* (1980).[42]

"There could be serious arms control negotiations, but only after we have built up our forces. . . . In ten years."

Paul Nitze in May 1981. Not long thereafter he was appointed to lead the U.S. negotiators in the Geneva talks on intermediate nuclear forces. The talks collapsed in December 1983.[43]

"If you think our proposal is one-sided, wait till you see theirs!"

Walter Slocombe, former Department of Defense official, commenting on the Reagan administration's "zero option" proposal in November 1981.[44]

"This stuff [arms control] is a soporific. It puts our society to sleep. It does violence to our ability to maintain adequate defenses."

Richard Perle, an official of the Department of Defense, at a meeting of Administration policymakers in January 1984.[45]

6

MONEY, SECRETS, AND SCIENTISTS

I have always heard it said that peace brings riches; riches bring pride; pride brings anger; anger brings war; war brings poverty; poverty brings humanity; humanity brings peace; peace, as I have said, brings riches, and so the world's affairs go round.

The Italian historian Luigi da Porto, 1509.[1]

"Pas d'argent, pas de Suisses."

"No money, no Swiss [mercenaries]." French saying of the sixteenth century.[2]

To wage war, you need first of all money; second, you need money, and third, you also need money.

Prince Montecuccolli of the Hapsburg court (1609–1680).[3]

Total War

Nowadays the whole art of war is reduced to money: and nowadays, that prince who can best find money to feed, cloath and pay his army, not he that has the most valiant troops, is surest of success and conquest.

Charles Davenant, *Essay Upon Ways and Means of Supplying the War* (1695).[4]

Wars are not paid for in wartime, the bill comes later.

Benjamin Franklin.[5]

"What the Governments will all come to see soon more or less clearly is that if they persist in squandering the resources of their people in order to prepare for a war which has already become impossible without suicide, they will only be preparing the triumph of a socialist revolution."

The Polish banker Ivan Bloch in conversation with the British journalist William T. Stead in the late 1890's.[6]

Just imagine—Millikan is said to have a hundred thousand dollars *a year* for his researches.

The German physicist Wilhelm Roentgen in a letter of November 13, 1921, remarking on the unprecedented resources of the new California Institute of Technology under Robert A. Millikan.[7]

Money, Secrets, and Scientists

Eisenhower: "Tell me, Mr. Khrushchev, how do you decide
    on funds for military expenditures? . . ."
Khrushchev: "Well, how is it with you?"
Eisenhower: "It's like this. My military leaders come to me
    and say, 'If we don't get the funds we need, we'll fall
    behind the Soviet Union.' So I invariably give in.
    That's how they wring money out of me. . . . Now
    tell me, how is it with you?"
Khrushchev: "It's just the same. . . ."
Eisenhower: "Yes. . . . You know, we really should come to
    some sort of an agreement in order to stop this
    fruitless, really wasteful rivalry."
Khrushchev: "That's one of our dreams."

> Khrushchev, recalling a private conversation at Camp David in 1959
> on "one point on which Eisenhower and I agreed."[8]

"I understand your joy. We felt it too not so long ago. But
you know, it's very expensive!"

> Khrushchev to de Gaulle during a visit to France after de Gaulle
> immediately shared the news of France's second successful atom
> bomb test on April 1, 1960.[9]

"Missiles will be able to do anything bombers can do—
cheaper."

> Secretary of Defense Robert McNamara, to associates in 1962.[10]

"There is a saying in Washington called the Golden Rule: he who has the gold, rules."

> Charles W. Snodgrass, a former congressional staff member, 1981.[11]

"If I had thought my coat knew my plans, I would take it off and burn it."

> Frederick II of Prussia (1712–1786).[12]

A secret is no secret when it is known to more than three persons.

> Sir Algernon West, one of Gladstone's secretaries.[13]

I dined with Bismarck alone . . . and then we talked and smoked. If you do not smoke under such circumstances, you look like a spy, taking down his conversation in your mind. Smoking in common puts him at his ease.

> Benjamin Disraeli, describing his visit with Bismarck in London in 1862.[14]

In the first place, too many people were concerned in the business of gathering information, and there was an ugly atmosphere of competition among them. . . . As soon as Intelligence found out anything, it proceeded to put it down on paper, mark the document in red ink "Top Secret," and then shut it away from all those likely to be interested in its contents, in a safe with a triple lock.

> The French historian Marc Bloch, on the intelligence failures of the French Army in 1940.[15]

"In fact Allen Dulles did come up to Mr. Mahon and Clarence Cannon and maybe one other congressman for half an hour on a Sunday afternoon in the basement of the Capitol and said, 'Mr. Mahon, I need X hundred million dollars for the CIA this year,' and they said, 'Fine, where do we sign the check,' and went home."

> Charles W. Snodgrass, former staff member of the Appropriations Committee responsible for intelligence, speaking in 1981.[16]

It is difficult, if not impossible, for anyone who has not been directly involved with the hidden conflict between rival intelligence bureaucracies to believe how bitter this conflict is, and the extreme forms it may take. . . . Anyone who believes that the rivalry between civilian and military intelligence agencies is any less intense today than it was between, say, comparable bureaucracies under the Nazi regime is either remarkably naive or has led a very sheltered bureaucratic life.

> Paul W. Blackstock, *The Strategy of Subversion* (1964).[17]

"All our estimates of the military balance are based on the assumption that Murphy's Law does not operate in the USSR."

> Herbert York, former director of Defense Research and Engineering at the Pentagon.[18]

"In the twenty-plus years I have spent in the intelligence field, it has been my experience that you almost never get, whether by human or technical collection, something that tells you the entire story about matters in which you are interested."

> Admiral Bobby R. Inman, former deputy director of the CIA, in a seminar at Harvard University, Spring 1981.[19]

"It is not a question of reluctance on the part of CIA officials to speak to us. Instead it is a question of our reluctance, if you will, to seek information and knowledge on subjects which I personally, as a member of Congress and as a citizen, would rather not have."

> Senator Leverett Saltonstall in Congress, April 9, 1956. Saltonstall was one of the Senate's chosen few to oversee the CIA.[20]

"If you believe in a program, you may have to break a little crockery to put it into effect."

> Allen Dulles, Director of Central Intelligence, 1953–1961.[21]

"At least we're getting the kind of experience we need for the next war."

> Allen Dulles, after a briefing by Harry Rositzke on the CIA's fruitless attempt to support anti-Soviet partisans in the Ukraine, 1953. [22]

The first question asked each defector was strictly prescribed: Did he have any information (called "early warning") of a pending attack on the United States or NATO?

> CIA field officer William Hood, on procedures in Vienna in the early 1950's. [23]

"I don't care what the C.I.A. does. All I want from them is twenty-four hours' notice of a Soviet attack."

> General George C. Marshall, while Secretary of Defense (1950–1951). [24]

"The publishers have to understand that we're never more than a miscalculation away from war and that there are things we're doing that we just can't talk about."

> President Kennedy, in conversation with his press secretary, Pierre Salinger, after refusing to discuss the Bay of Pigs debacle at a press conference on April 21, 1961. [25]

Total War

I quickly learned that controlling the newsbreaks, whatever the source, is the best entrée to the great men, who like to be up to date even more than they like to be well briefed.

> Ray Cline, former deputy director for intelligence, CIA.[26]

"I have accused analysts from time to time of writing at the highest possible classification level with as many codewords as possible, in the belief that they were more likely to get their work read at the highest levels. They usually deny it."

> Admiral Bobby Inman, speaking in 1981.[27]

"Nobody works for us for ideological motives anymore."

> Chief of the Czech intelligence service in Bonn, at a meeting in the Soviet Embassy in 1976.[28]

Once one gets a taste for it, it's hard to drop.

> Allen Dulles in a letter to Robert Murphy in 1951, explaining his fascination with intelligence.[29]

The crowd is unable to digest scientific facts, which it scorns and misuses to its own detriment and that of the wise. Let not pearls, then, be thrown to swine.

> Roger Bacon (1214–1292), explaining why he hid his formula for gunpowder in a cryptogram.[30]

Money, Secrets, and Scientists

If the two countries or governments are at war, the men of
science are not—that would indeed be a civil war of the worst
sort.

> The British chemist Sir Humphry Davy in 1807, explaining why he
> had accepted an award from Napoleon, with whom Britain was at
> war.[31]

"If you'd give less attention to those scientific things
and more to your naval gunnery, there might come a time
when you would know enough to be of some use to your
country."

> The superintendent of the United States Naval Academy to Albert
> Michelson, the first American to win a Nobel Prize in physics, at his
> graduation from the academy in 1873.[32]

I hate and fear "science" because of my conviction that; for
long to come if not for ever; it will be the remorseless enemy
of mankind. . . . I see it darkening men's minds and harden-
ing their hearts; I see it bringing a time of vast conflicts, which
will pale into insignificance "the thousand wars of old," and,
as likely as not, will whelm all the laborious advances of
mankind in blood-drenched chaos.

> George Gissing, *The Private Papers of Henry Ryecroft* (1903).[33]

97

Total War

We must not prepare poisonous gases or debase science through similar misuse; but we should give our soldiers and sailors every legitimate aid and every means of protection.

> George Ellery Hale, director of Mount Wilson Observatory and a founder of the National Academy of Sciences, to President Woodrow Wilson in July 1916.[34]

Those who shared in the consciousness of the University's power and resourcefulness can never be fully content to return to the old routine of the days before the war.

> Dean Frederick J. E. Woodbridge of Columbia University in 1919.[35]

"I want to inform you that without the Jews there is no mathematics and physics possible in Germany."

> The physicist Max Planck to Hitler in 1933, protesting the expulsion of Jews from German universities.[36]

"No, I don't want to leave. Germany needs me."

> Werner Heisenberg, a leader of the German atom bomb program during World War II, when the American physicist Samuel Goudsmit asked him in May 1945 if he would like to go to the United States.[37]

98

In my youthful way I wondered, "If only some day a hundred years from now, a little street or even an alley could be named after me."

Stanislaw Ulam, co-inventor, with Edward Teller, of the hydrogen bomb, describing his delight on observing that lecture halls and even streets in Paris had been named after great mathematicians.[38]

"Well, Edward, now that you have your H-bomb, why don't you use it to end the war in Korea?"

Oppenheimer remark at lunch with Teller and Rabi in November 1952, following the first successful test of a thermonuclear device or H-bomb, as recalled by Teller.[39]

"You go right ahead and try, Doctor."

Senator J. William Fulbright at a hearing on anti-ballistic missiles, after the analyst Donald Brennan said the subject was too complex to explain to laymen, 1969.[40]

7

THE NATURE OF WAR

In the life of a society models of weapons change very often, models of tools change less often and social institutions very seldom, while religious institutions continue unchanged for millennia.

André Leroi-Gourhan, a French authority on neolithic cave art.[1]

How, O Sumer, are thy mighty fallen!
The holy king is banished from his temple
The temple itself is destroyed, the city demolished
The leaders of the nation have been carried off into captivity
A whole empire has been overthrown by the will of the gods.

"Lamentation over the Destruction of Ur," ancient Sumerian poem on the fate of Ibi-Sin, the last king of Ur (2029–2006 B.C.).[2]

The chief reason why Asiatics are less warlike and more gentle in character than Europeans is the uniformity of the seasons,

101

which show no violent changes. . . . For there occur no mental shocks . . . which are more likely to steel the temper and impart to it a fierce passion than is a monotonous sameness. For it is changes of all things that rouse the temper of man and prevent its stagnation. For these reasons, I think, Asians are feeble. Their institutions are a contributory cause, the greater part of Asia being governed by kings. Now where men are not their own masters and independent . . . they are not keen on military efficiency. . . . All their worthy, brave deeds merely serve to aggrandize . . . their lords, while the harvest they themselves reap is danger and death. . . . [Yet] all the inhabitants of Asia, whether Greek or non-Greek, who are not ruled by despots, but are independent, toiling for their own advantage, are the most warlike of all men. For it is for their own sakes that they run their risks, and in their own persons do they receive the prizes of their valour as likewise the penalty of their cowardice.

Hippocrates (ca. 460–377 B.C.), *Airs Waters Places*.[3]

For to win one hundred victories in one hundred battles is not the acme of skill. To subdue the enemy without fighting is the acme of skill. Thus, what is of supreme importance in war is to attack the enemy's strategy. Next best is to disrupt his alliances. The next best is to attack his army. The worst policy is to attack cities. Attack cities only when there is no alternative. . . . Thus, those skilled in war subdue the enemy's army without battle. They capture his cities without assaulting them and overthrow his state without prolonged operations.

Sun Tzu, Chinese military writer of the fourth century B.C., and author of *The Art of War*.[4]

The Nature of War

"Memini me Alexandrum esse, non mercatorum."

["I remember that I am Alexander and not a merchant."]

> Alexander the Great (356–323 B.C.). The quotation from his biographer, Quintus Curtius, was a favorite of Charles XII of Sweden.[5]

The wild animals that range over Italy have a hole, and each of them has its lair and nest, but the men who fight and die for Italy have no part or lot in anything but the air and the sunlight.

> Roman political leader Tiberius Gracchus, 133 B.C.[6]

Do not all peoples hate wrongdoing? Yet, is it not rampant among them all? Are not the praises of truth sung by all nations? Yet is there a single race or tribe that really adheres to it? What nation likes to be oppressed by a stronger power? Or who wants his property plundered unjustly? Yet, is there a single nation that has not oppressed its neighbor? Or where in the world will you find a people that has not plundered the property of another?

> "The Coming Doom," one of the documents known as the Dead Sea Scrolls, written by the Qumran community in the first century B.C.[7]

Total War

Indeed there is nothing more stable, more prosperous, more glorious than a state which abounds in trained soldiers. For it is not the splendour of uniforms nor the abundance of gold, silver and precious stones which make us respected or sought out by our enemies: they can only be forced to submit through terror.

> The fourth-century A.D. Roman military writer Vegetius.[8]

We see that the Roman people have conquered the world by nothing other than drill in arms.

> Vegetius.[9]

Lamachus rebuked an officer for a mistake. When the officer said that he would not do it a second time Lamachus said: In war there is no second time.

> Plutarch, first century A.D.[10]

"Solitudinem faciunt pacem appellant."
> ["They make a desert and call it peace."]

> A Briton of the first century A.D., speaking of the Romans, as quoted by Tacitus, *Agricola,* 30 (A.D. 98).[11]

The Nature of War

"Gifts to friends, steel to foes."

Emperor Marcian to Attila the Hun, A.D. 450.[12]

"The greatest pleasure is to vanquish your enemies and chase them before you, to rob them of their wealth and see those dear to them bathed in tears, to ride their horses and clasp to your bosom their wives and daughters."

Genghis Khan (1162–1227).[13]

Houses and churches no longer presented a smiling appearance with newly thatched roofs but rather the lamentable spectacle of scattered smoking ruins amid nettles and thistles springing up on every side. The pleasant sound of bells was heard indeed, not as a summons to divine worship but as a warning of hostile intention, so that men might seek out hiding places while the enemy were still on the way. What more can I say?

Jean de Venette, a Carmelite friar, on fourteenth-century France during the Hundred Years' War.[14]

"Do you not know that I live by war and that peace would be my undoing?"

Sir John Hawkwood, a fourteenth-century mercenary.[15]

Total War

"Know what you fight for and love what you know."

Oliver Cromwell (1599–1658).[16]

"They now *ring* the bells, but they will soon *wring* their hands."

British Prime Minister Sir Robert Walpole, on the eve of Britain's war with Spain in 1739.[17]

"Alas! Alas! how this unmeaning stuff spoils all my comfort in my friends' conversation! Will the people never have done with it; and shall I never hear a sentence again without the *French* in it? Here is no invasion coming, and you *know* there is none. . . .*Oh, pray* let us hear no more of it!"

Samuel Johnson in conversation with his friend Mrs. Thrale.[18]

May we never see another war, for in my opinion *there never was a good war, or a bad peace.*

Benjamin Franklin in a letter to Josiah Quincy, September 11, 1783.[19]

War is a violent condition. One should make it *à l'outrance* [to the utmost] or go home.

Lazare Carnot, military leader during the French Revolution.[20]

The Nature of War

"In war it isn't so much what one does that matters, but that whatever action is agreed upon should be carried out with unity and energy."

> Gerhard von Scharnhorst, the Duke of Brunswick's chief of staff, during the catastrophic campaign which ended in Prussia's defeat by Napoleon in 1806, later quoted by Clausewitz.[21]

It was my ambition to write a book that would not be forgotten after two or three years, and that possibly might be picked up more than once by those who are interested in the subject.

> Carl von Clausewitz, Prussian general and military writer, in a note discovered by his wife after his death in 1831.[22]

We are not interested in generals who win victories without bloodshed. The fact that slaughter is a horrifying spectacle must make us take war more seriously, but not provide an excuse for gradually blunting our swords in the name of humanity. Sooner or later someone will come along with a sharp sword and hack off our arms.

> Carl von Clausewitz, *On War* (1832).[23]

But no less practical is the importance of another point that must be made absolutely clear, namely that *war is nothing but the continuation of policy with other means.*

> Carl von Clausewitz in a note of July 10, 1827, on his plans for revising *On War.*[24]

107

Total War

When a war has at length by its long continuance roused the whole community from their peacetime occupations and brought all their petty undertakings to ruin, it will happen that those very passions which once made them value peace so highly become directed into war. War, having destroyed every industry, in the end becomes itself the one great industry, and every eager and ambitious desire sprung from equality is focused on it.

Alexis de Tocqueville, *Democracy in America* (1836).[25]

Conquest is the premium given by nature to those national characters which their national customs have made most fit to win in war, and in most material respects those winning characters are really the best characters. The characters which do win in war are the characters which we should wish to win in war.

English economist and journalist Walter Bagehot (1826–1877).[26]

"This man Wellington is so stupid he does not know when he is beaten and goes on fighting."

Napoleon at Waterloo.[27]

"Sir, my strategy is one against ten, my tactics ten against one."

Duke of Wellington to Blücher before Waterloo in 1815.[28]

The Nature of War

No man can properly command an army from the rear. . . . Some men think that modern armies may be so regulated that a general can sit in an office and play on his several columns as on the keys of a piano; this is a fearful mistake. The directing mind must be at the very head of an army. . . . Every attempt to make war easy and safe will result in humiliation and disaster.

> General William Tecumseh Sherman in his memoirs of the Civil War (1861–1865). Sherman traveled light, kept his staff small, and never shirked danger.[29]

The almost entire separation of the staff from the line, as now practised by us, has proved mischievous, and the great retinues of staff-officers with which some of our earlier generals began the war were simply ridiculous. . . . A bulky staff implies a division of responsibility, slowness of action, and indecision, whereas a small staff implies activity and concentration of purpose.

> General Sherman in his memoirs.[30]

No plan survives contact with the enemy.

> Helmuth von Moltke the elder, Prussian field marshal (1800–1891).[31]

To new soldiers the sight of blood and death always has a sickening effect, but soon men become accustomed to it, and

Total War

I have heard them exclaim on seeing a dead comrade borne to the rear, "Well, Bill has turned up *his* toes to the daisies."

General Sherman, commander of Union forces on the infamous march through Georgia in 1864, one of the concluding campaigns of the Civil War.[32]

I had heard so much of the cannon fever, that I wanted to know what kind of thing it was. . . . I had now arrived quite in the region where the balls were playing across me: the sound of them is curious enough, as if it were composed of the humming of tops, the gurgling of water, and the whistling of birds. . . . In the midst of these circumstances, I was soon able to remark that something unusual was taking place within me. . . . It appeared as if you were in some extremely hot place, and, at the same time, quite penetrated by the heat of it, so that you feel yourself, as it were, quite one with the element in which you are. The eyes lose nothing of their strength and clearness; but it is as if the world had a kind of brown-red tint.

Goethe, describing the Battle of Valmy in 1792.[33]

"The god of war."

Stalin's definition of artillery, according to the Soviet arms negotiator V. S. Semenov in the fall of 1970.[34]

The Nature of War

"Why, you take the most gallant sailor, the most intrepid airman, or the most audacious soldier, put them at a table together—what do you get? *The sum of their fears.*"

Churchill, on the Chiefs of Staffs system, during lunch on board the *Renown*, November 16, 1943, en route to the Cairo Conference.[35]

I begin to regard the death and mangling of a couple thousand men as a small affair, a kind of morning dash.

General Sherman to his wife, Ellen, in a letter dated June 30, 1864.[36]

"The men that war does not kill it leaves completely transparent."

Colonel Castelo Branco of Brazil to Vernon Walters as they sat through a heavy artillery bombardment on the Italian front in World War II.[37]

If the people raise a howl against my barbarity and cruelty, I will answer that war is war, and not popularity-seeking. If they want peace, they and their relatives must stop the war.

General Sherman in a letter to General Halleck on September 4, 1864, justifying his scorched earth policy.[38]

111

Total War

The main thing in true strategy is simply this: first deal as hard blows at the enemy's soldiers as possible, and then cause so much suffering to the inhabitants of a country that they will long for peace and press their Government to make it. Nothing should be left to the people but eyes to lament the war.

> General Philip Henry Sheridan (1831–1888).[39]

"My dear child, why all this noise? What is it all about?"

> French mother to her soldier son just behind the line during a battle in 1918.[40]

"The great questions of the day will be decided not by speeches and majority votes . . . but by iron and blood."

> Otto von Bismarck, on September 24, 1862, shortly after taking office as German chancellor. The phrase was frequently quoted thereafter with the last two words reversed—"blood and iron"— perhaps because they sounded better this way in English.[41]

In our Country . . . one class of men makes war and leaves another to fight it out.

> General Sherman to General O. O. Howard, May 17, 1865.[42]

The politician should fall silent the moment that mobilization begins.

> Helmuth von Moltke, chief of the Prussian general staff, 1858–1888.[43]

"It is well that war is so terrible—we would grow too fond of it."

> General Robert E. Lee to a staff member, while observing columns of men advancing under fire.[44]

"Well, I've been a war correspondent long enough to have the right to say I like it. There's a thrill about it that's pleasant."

> Richard Harding Davis, describing the experience of being under fire on the Bulgarian front in the winter of 1914–15.[45]

No lesson seems to be so deeply inculcated by the experience of life as that you never should trust experts. If you believe the doctors, nothing is wholesome: if you believe the theologians, nothing is innocent: if you believe the soldiers, nothing is safe.

> Lord Salisbury (1830–1903) in a letter to Lord Lytton, June 15, 1877.[46]

Total War

The rifle, effective as it is, cannot replace the effect produced by the speed of the horse, the magnetism of the charge, and the terror of cold steel.

British Army *Cavalry Training Manual* (1907).[47]

Without War the world would become swamped in materialism.

Helmuth von Moltke, 1880.[48]

Experience proves that the man who obstructs a war in which his nation is engaged, no matter whether right or wrong, occupies no enviable place in life or history. Better for him, individually, to advocate "war, pestilence, and famine," than to act as obstructionist to a war already begun.

Ulysses S. Grant in his memoirs, 1885.[49]

If men are mad enough they will fight. If not, the ordinary means of diplomacy will do.

The pioneer American sociologist William Graham Sumner (1840–1910).[50]

"Aleksey Nikolayevich, you are not familiar with Russia's internal situation. *We need a little victorious war to stem the tide of revolution.*"

> V. K. Plehve, Russian minister of the interior, to General Aleksey Kuropatkin in the early days of the disastrous Russo-Japanese War, 1904–1905.[51]

Within a year from the breaking of Germany's power . . . our Imperialists will be calling out for a strong Germany to balance a threatening Russia.

> Henry Noel Brailsford of the United Kingdom, on British war aims in World War I.[52]

"England has neither eternal friends nor eternal enemies. She has only eternal interests."

> The British prime minister Lord Palmerston (1784–1865).[53]

He that commands the sea is at great liberty and may take as much or as little of the war as he will.

> Francis Bacon (1561–1626).[54]

Total War

When one gets a close view of influential people—their bad relations with each other, their conflicting ambitions, all the slander and the hatred—one must always bear in mind that it is certainly much worse on the other side, among the French, English, and Russians, or one might well be nervous. . . . The race for power and personal positions seems to destroy all men's characters. I believe that the only creature who can keep his honor is a man living on his own estate; he has no need to intrigue and struggle—for it is no good intriguing for fine weather.

General Max Hoffmann of the German High Command in World War I.[55]

In some happy corners of the earth, they say, where nature brings forth abundantly whatever man desires, there flourish races whose lives go gently by, unknowing of aggression or constraint. This I can hardly credit; I would like further details about these happy folk. The Bolshevists, too, aspire to do away with human aggressiveness by ensuring the satisfaction of material needs and enforcing equality between man and man. To me this hope seems vain. Meanwhile they busily perfect their armaments, and their hatred of outsiders is not the least of the factors of cohesion amongst themselves. In any case . . . complete suppression of man's aggressive tendencies is not in issue; what we may try is to divert it into a channel other than warfare.

Sigmund Freud, "Why War?" (1933).[56]

The Nature of War

Restricted warfare was one of the loftiest achievements of the eighteenth century. It belongs to the class of hot-house plants which can only thrive in an aristocratic and qualitative civilization. We are no longer capable of it. It is one of the fine things we have lost as a result of the French Revolution.

Guglielmo Ferrero, *Peace and War* (1933).[57]

"You big nations are hard; you talk of our honor, but you are far away."

Prince Paul of Yugoslavia to U.S. Ambassador Arthur Lane, who had been urging him to resist Hitler. A week later, on March 27, 1941, the Prince Regent was ousted and he fled to Greece.[58]

"This war is not as in the past; whoever occupies a territory also imposes on it his own social system. Everyone imposes his own system as far as his army can reach. It cannot be otherwise."

Stalin to Milovan Djilas in April 1945.[59]

War does not begin to reveal its malignity till the war-making society has begun to increase its economic ability to exploit Physical Nature and its political ability to organize 'manpower'; but, as soon as this happens, the God of War to which the growing society has long since been dedicated proves himself a Moloch.

Arnold Toynbee, *War and Civilization* (1950).[60]

"The principal cause of war is war itself."

The American sociologist C. Wright Mills, *The Causes of World War Three* (1958).[61]

Putting aside all the fancy words and academic doubletalk, the basic reason for having a military is to do two jobs—to kill people and to destroy the works of man.

General Thomas S. Power, *Design for Survival* (1964).[62]

The stakes of war are the existence, the creation or the elimination of states.

Raymond Aron, *Peace and War* (1966).[63]

Sweeping judgments, malicious gossip, inaccurate statements which spread a misleading impression—these are symptoms of the moral and mental recklessness that gives rise to wars.

B. H. Liddell Hart, *Why Don't We Learn from History?* (1971).[64]

The Nature of War

"There are always three choices—war, surrender, and present policy."

Henry Kissinger, on the strategy of writing memos.[65]

Conventional military men think of combat psywar almost exclusively in terms of leaflets or broadcasts appealing to the enemy to surrender. Early on, I realized that psywar had a wider potential than that. A whole new approach opens up, for example, when one thinks of psywar in terms of playing a practical joke.

Major General Edward Lansdale, *In the Midst of Wars* (1972). Lansdale put his theory into practice in the Philippines in the 1940's, and in Vietnam in the 1950's.[66]

"What fun is it to be a four-star general, head of CINCPAC, if any time a real war starts Lyndon Johnson goes over the bombing list every night and tells you what you can or can't bomb?"

Charles W. Snodgrass, a longtime civil servant in the national security community, describing recent changes in military communications in 1981.[67]

119

Total War

We would be blind therefore if we did not recognize that the causes which have produced war in the past are operating in our own day as powerfully as at any time in history.

Michael Howard, *The Causes of Wars* (1983).[68]

If you want peace, *understand* war.

Motto of the British military writer B. H. Liddell Hart.[69]

8

THE CENTURY OF TOTAL WAR

We have not had hostilities with either France, America, or Russia; and when not at war with any of our peers, we feel ourselves to be substantially at peace.

The English historian Sir Edward Shepherd Creasy, 1851.[1]

Modern wars have not materially changed the relative values or proportions of the several arms of service. . . . If any thing, the infantry has been increased in value. . . . All great wars will, as heretofore, depend chiefly on the infantry. . . . Earth-forts, and especially field-works, will hereafter play an important part in wars, because they enable a minor force to hold a superior one in check for a *time*, and time is a most valuable element in all wars.

General William T. Sherman's words were prophetic of trench fighting in World War I but were ignored.[2]

Total War

The continent has been converted into a series of gigantic camps, within each of which a whole nation stands in arms.

> *The Economist,* 1879, describing the European response to Prussia's defeat of France in 1870.[3]

It is virtually impossible that anyone can have accurately pictured to himself the scene in its fulness which the next great battle will present to a bewildered and shuddering world.

> The British journalist Archibald Forbes, in a volume of memoirs (1895) describing his experiences in the Franco-Prussian War of 1870 and the Russo-Turkish War of 1878.[4]

"Some damned foolish thing in the Balkans."

> Bismarck, after being asked what would set off the next general European war.[5]

"I wish you to kill and burn. The more you burn and kill the better it will please me."

> General Jake Smith of the U.S. during the Philippine insurrection, 1899–1901.[6]

For twenty-two years, since the Congress of Berlin [in 1878], Europe . . . has been at peace. This peace . . . has been unquiet, apprehensive; the outcome, not of the removal of the causes of strife, but of the very perfection of the preparations for a struggle which, regarded as inevitable, is ever postponed, because no one can measure its horror nor forecast its result. This 'armed peace,' the legacy of Bismarck and of Moltke to the world, has been for two decades the dominant fact in European international politics. Amid constant rumours of war the temper of responsible statesmen has never been less warlike; for no Foreign Secretary would now dare, like Palmerston, to end a despatch with a glib threat of extreme consequences, unless he were backed by an overwhelming public opinion. And public opinion is no longer likely to declare lightly for war in countries where every man is liable to military service. And so, in spite of national rivalries, which have never been more intense, in spite of unhealed sores and unsatisfied ambitions, the peace of Europe remains, founded upon fear.

W. Alison Phillips, *Modern Europe* (1901).[7]

There is only one limit possible to the war preparations of a modern European state; that is, the last man and the last dollar it can control. What will become of the mixture of sentimental social philosophy and warlike policy? There is only one thing rationally to be expected, and that is a frightful effusion of blood in revolution and war during the century now opening.

William Graham Sumner in "War," an essay first published in 1903.[8]

Total War

From a European war a revolution may spring up and the ruling classes would do well to think of this. But it may also result, over a long period, in a crisis of revolution, of furious reaction, of exasperated nationalism, of stifling dictatorship, of monstrous militarism, a long chain of retrograde violence.

> The French politician Jean Jaurès in 1905 when war between France and Germany seemed close. It is not often that a flat prediction comes closer to the mark. It *all* happened—revolution and counter-revolution alike.[9]

"These two great peoples have nothing to fight about, no prize to fight for, and no place to fight in."

> Winston Churchill on the friendship between Britain and Germany, in an election speech in Swansea, August 14, 1908.[10]

"A single British soldier—and we will see to it that he is killed."

> General Ferdinand Foch in January 1910, answering General Henry Wilson, who had asked what was the smallest British force that would be of use in a new war.[11]

"During the past few days, I have given a great deal of thought to this avalanche that is carrying us toward war. At all

124

costs we must try to open our Emperor's eyes to the danger. . . . Someone—someone among us—must have the courage, as in Andersen's fairy tale, to tell the Emperor that he has no clothes!"

> Baron Taube in the spring of 1914, reporting the concern at Russia's misplaced confidence it was ready for a general European war.[12]

"We all muddled into war."

> British Prime Minister David Lloyd George, on World War I.[13]

"If you are speaking of guilt for this war, we must be honest enough to admit that we also have our own share in it. It would be an understatement to say that I am weighed down by this thought; it never leaves me; I live with it."

> Bethmann-Hollweg to the German journalist Theodor Wolff on February 9, 1915.[14]

Black and hideous to me is the tragedy that gathers, and I'm sick beyond cure to have lived on to see it. You and I, the ornaments of our generation, should have been spared this wreck of our belief that through the long years we had seen civilization grow and the worst become impossible. The tide that bore us along was then all the while moving to *this* as its grand Niagara—yet what a blessing we didn't know it. It seems to me to *undo* everything, everything that was ours, in the most horrible retroactive way—but I avert my face from the monstrous scene!

> Henry James in a letter to a friend after the outbreak of war in August 1914. The following year he became a British subject as a mark of loyalty to the country where he had lived since 1876. [15]

"Even if we end in ruin, it was beautiful."

> General Erich von Falkenhayn, the Prussian Minister of War, on August 4, 1914, the first day of World War II. [16]

"Gentlemen: when the barrage lifts!"

> A British toast, reprinted in the London *Times* on the anniversary of the Battle of the Somme. [17]

I can still hear the awful lamentation of the women and the drunken uproar of the men during the first days of war.

> General Petro Grigorenko of Russia, recalling the outbreak of World War I, when he was seven years old. [18]

The Century of Total War

"I suppose there is no fighting on Sundays."

> The American painter John Singer Sargent (1856–1925) on reaching the front in World War I. From drawings there, he painted a twenty-foot-long canvas of blindfolded survivors of the battle of Ypres, "Gassed," often called his magnum opus.[19]

Smith with homely truth observed: "You can tell a shell hole by its contents, two Mills bombs and a lump of faeces."

> Guy Chapman, in his memoir of the Western Front in World War I, *A Passionate Prodigality* (1933).[20]

The conclusion seems to be that we are to-day entering on an era in which war will not only flourish as vigorously as in the past, although not in so chronic a form, but with an altogether new ferocity and ruthlessness, with a vastly increased power of destruction, and on a scale of extent and intensity involving an injury to civilisation and humanity which no wars of the past ever perpetrated. . . . It seems clear that we have to recognise that our intellectual leaders of old who declared that to ensure the disappearance of war we have but to sit still and fold our hands while we watch the beneficent growth of science and intellect were grievously mistaken.

> Havelock Ellis, *Essays in War-time* (1917).[21]

There is no doubt in my mind that he has no drive, that he is almost a beaten man, that he is always turning to a Peace to

127

get him out of his difficulties—he spoke to me of peace 2 or 3 times again to-day—and that I really begin to think that he had better be relieved.

> General Henry Wilson in his diary, describing the British commander in chief on the Western Front, Field Marshal Douglas Haig, at the height of Germany's last major offensive of the war, April 9, 1918.[22]

There is nothing like it on earth, nor can be. . . . There are miles upon miles of flat, empty, broken, and tumbled stone quarry. . . . Every ruler, leading statesman, or president of a republic ought to be brought to see it. . . . Then there would be no more wars.

> Rudolf Binding, a lieutenant in the German army, describing the wasteland north of Moreuil Ridge; in his diary, April 4, 1918.[23]

"I suppose our campaigns are ended, but what an enormous difference a few days would have made."

> General John "Black Jack" Pershing, on the morning of the armistice, November 11, 1918.[24]

Because good Europeans hate war in 1926 it does not follow that they hated war in 1914 or that they will hate it in 1964.

> General Sir Ian Hamilton, commander of British forces at Gallipoli, in the thirteenth edition of *The Encyclopedia Britannica* (1926).[25]

"I think it is well for the man in the street to realize that there is no power on earth that can protect him from being bombed. Whatever people may tell him, the bomber will always get through. . . . I just mention that . . . so that people may realize what is waiting for them when the next war comes."

British statesman Stanley Baldwin in the House of Commons on November 10, 1932.[26]

"*Quatsch!* [Nonsense!] I have seen dictatorship in Russia. In Germany it just couldn't happen."

The physicist Wolfgang Pauli in conversation with Edward Teller and Walter Heitler in Göttingen the night after the Reichstag Fire, February 1933.[27]

There is only one purpose to which a whole society can be directed by a deliberate plan. That purpose is war, and there is no other.

Walter Lippmann, *The Good Society* (1936).[28]

Every technological improvement applied to the machinery of destruction tightens the grip which modern war has on the common man's life. The scope of war has become as large as that of peace, or indeed even larger, since under modern conditions it is the interest of efficient war to militarize peace.

Hans Speier and Alfred Kähler, *War in Our Time* (1939).[29]

"Go now and do likewise to the Hun."

British Squadron Commander Bill Staton during a briefing of crews before a raid on Germany in 1940, stabbing a finger at photos on the wall of bomb damage in London, Coventry, and Southampton.[30]

Opposite me sat a woman refugee who had been driven from a village in the East. Her mind was occupied during the day, but while sitting at leisure in the dark [in an air raid shelter in Germany in 1945], her hands in her lap, grief over the abandoned house took hold of her. Unaware of who was listening, she enumerated her losses in a moaning soliloquy, found words of endearment for cows and hens, for the fruit trees in the garden, for her bees and flowers. She recited and repeated the list of her former possessions in a haunting chant while rocking herself backward and forward in an agony of home-sickness.

Lali Horstmann, *We Chose to Stay* (1954).[31]

We should never allow ourselves to apologize for what we did to Germany.

Churchill, in a note to a former staff officer of Bomber Command after World War II.[32]

"It is certain in my opinion that Europe would have been communized and London would have been under bombardment some time ago, but for the deterrent of the atomic bomb in the hands of the United States."

Winston Churchill in March 1949.[33]

"There are an increasing number of signs of toughness on the part of the Kremlin; the informal opinion of the Joint Chiefs now is that the Soviet Union could begin a major attack from a standing start so that the usual signs of mobilization and preparation would be lacking; there are increasing indications that some of the basic elements of Communist dogma no longer hold, i.e., that the Communist bastion has infinite time in which to achieve its purpose, that capitalist nations carry within themselves the seeds of their own destruction which require watering but not planting by the Soviet Union, that the Red Army is used only when a revolutionary atmosphere makes the situation right for the coup de grace, etc."

Diplomat Paul Nitze, at a meeting of the State Department Policy Planning Staff, February, 2, 1950.[34]

The year of maximum danger.

Paul Nitze's phrase for 1954, writing of the threat posed by Soviet nuclear weapons in National Security Council Memorandum Number 68 (1950), one of the basic documents of the Cold War.[35]

Total War

"We can't afford to fight limited wars. We can only afford to fight a big war, and if there is one that is the kind it will be."

> Secretary of Defense Charles E. Wilson in the 1950s.[36]

Lewis Strauss: "An H-bomb . . . can be made to be as large as you wish . . . large enough to take out a city."
Reporter: "How big a city?"
Strauss: "Any city."

> Strauss, chairman of the Atomic Energy Commission, following a presidential news conference on March 31, 1954, on being asked how large and powerful an H-bomb could be. On the way back to the White House, Eisenhower commented, "Lewis, I wouldn't have answered that one that way."[37]

"In the conduct of foreign affairs, we do so many things we can't explain that, once in a while, something happens. . . . There is a very great aggressiveness on our side that you have not known about. . . . Here's the thing to remember: suppose one day, we get into a war. If too many people knew we had done something provocative—I just want to say that we *might* have to answer to charges of being too provocative rather than being too sweet."

> President Eisenhower in conversation with Senator William Knowland, November 24, 1954, explaining why he did not want to break off diplomatic relations with the Soviet Union after they shot down an American B-29 on November 7. The plane was on a reconnaissance mission and may have flown over Soviet territory, something Knowland did not know.[38]

The critical number of U.S. targets is not large. It may be less than 50. We believe that 200 nuclear bombs of megaton and kiloton yield, if delivered on selective targets with practical accuracy, could decisively defeat us, and that a first attack could be fatal if we were surprised and unprepared.

> Report of the Technological Capabilities Panel of the Science Advisory Committee, referred to as the Killian Committee after its chairman, James Killian, February 14, 1955.[39]

We believe that a small-scale surprise attack with relatively few nuclear bombs could do severe damage to the United States. At the same time we also believe that the United States needs a large offensive striking force capable of hitting many targets with megaton bombs. This looks like a contradiction. . . .

> Report of the Technological Capabilities Panel of the Science Advisory Committee (Killian Committee), February 14, 1955.[40]

"I'm leaving because I found myself making decisions from fatigue."

> Charles E. Wilson, explaining why he resigned as Secretary of Defense, August 1957.[41]

"You can't have this kind of war. There just aren't enough bulldozers to scrape the bodies off the streets."

> President Eisenhower, speaking to the panel which had just given him the Gaither report, "Deterrence and Survival in the Nuclear Age," November 1957.[42]

"But the real problem is not one of weapons; it's one of peace or war. The situation is a highly dangerous one, and I think that the people with the strongest nerves will be the winners. That is the most important consideration in the power struggle of our time. The people with weak nerves will go to the wall."

> Khrushchev to Egyptian President Nasser in July 1958.[43]

"Alliances are held together by fear, not by love."

> Harold Macmillan to Nixon in 1959.[44]

"Bill, just remember this: if the United States is willing to go to war over Berlin, there won't be a war over Berlin."

> John Foster Dulles to aide William Macomber from his hospital bed where he was dying of cancer, May 1969.[45]

"Every time you lift the phone, Mr. President, I think you may say that you intend to go, and I wonder what answer I would give."

> Harold Macmillan to Kennedy during a meeting in Bermuda, December 21, 1961.[46]

"We are going to insist that the Germans stop talking about a finger on the trigger without a plan to use it. We are going to make them resolve the question of whose finger, and when it is to push."

> McNamara to John McNaughton, Pentagon official, after a long wrangle with the Germans about joint control of nuclear weapons in 1966.[47]

"I did it [halted the bombing of North Vietnam] twelve times, and not a one of them did a damn bit of good. Ike was different. The Russians feared Ike. They didn't fear me."

> Lyndon Johnson to Richard Nixon in 1969.[48]

One could argue—and many did—that nations whose combined populations and Gross National Products were at least three times the Soviet Union's should be able to mount a conventional defense against the Warsaw Pact. The difficulty was that no member of the [NATO] Alliance, including the United States, was prepared to make the effort.

> Henry Kissinger.[49]

"I don't see why we have to let a country go Marxist just because its people are irresponsible."

> Kissinger at a secret meeting in June 1970, on the refusal of the right in Chile to unite against Allende in the election that year. Kissinger was arguing for a covert attempt by the CIA to block Allende's election.[50]

Total War

We are living in a pre-war and not a post-war world.

> Eugene Rostow, who later became Reagan's first director of the
> Arms Control and Disarmament Agency, in reply to a letter of May
> 1976.[51]

"But President Carter opened up his decision handbook, he really got into the procedures, ran through numerous scenarios and became very comfortable with it. He wanted to be able to be awakened at three o'clock in the morning and not be confused, and understand what he was going to have to see, or what he was about to hear, what the voice would sound like on the other end of the line, and that sort of thing."

> General William Odom, military adviser to the National Security
> Council, describing the 1977 Command Post Exercises (CPXs) with
> which Carter trained himself in the plans and procedures for waging
> a nuclear war.[52]

9

FOREBODINGS

"Let us cease to consider what, perhaps, may never happen, and what, when it shall happen, will laugh at human speculation."

Samuel Johnson, *The History of Rasselas, Prince of Abyssinia* (1759).[1]

"Sir, there is a great deal of ruin in a nation."

Adam Smith (1723–1790) to a friend worried that England would ruin herself in wars with France.[2]

It is the last hour: Pestilence and sword are raging in the world. Nation is rising against nation, the whole fabric of things is being shaken.

Gregory the Great to the Patriarch of Constantinople in A.D. 595.[3]

Total War

Some day science may have the existence of mankind in its power, and the human race commit suicide by blowing up the world.

Henry Adams in a letter, 1862.[4]

It is quite possible for such a state of feeling to exist between two States that a very trifling political motive for war may produce an effect quite disproportionate—in fact a perfect explosion.

Clausewitz, *On War*.[5]

If the war which has hung over our heads, like the sword of Damocles, for more than ten years past, ever breaks out, its duration and end cannot be foreseen. The greatest powers of Europe, armed as never before, will then stand face to face. No one power can be shattered in one or two campaigns so completely as to confess itself beaten, and conclude peace on hard terms. It may be a Seven Years' War; it may be a Thirty Years' War—woe to him who first sets fire to Europe.

Helmuth von Moltke the elder (1800–1891), writing in 1890.[6]

"The lamps are going out all over Europe; we shall not see them lit again in our lifetime."

Sir Edward Grey to a friend as they stood at a window watching the street lamps of London being lit on the evening of August 3, 1914, the outbreak of World War I.[7]

Forebodings

"What a world we live in and how the public would stare if they could look into our minds and our letter bags."

> Austen Chamberlain, British statesman, in 1915.[8]

It is useless to delude ourselves. All the restrictions, all the international agreements made during peacetime are fated to be swept away like dried leaves on the winds of war.

> The Italian theorist of air power and strategic bombing, General Giulio Douhet, 1928.[9]

"My God, we simply have to figure a way out of this situation. There's just no point in talking about 'winning' a nuclear war."

> President Eisenhower, in conversation in October 1956.[10]

Recently, he told me that not once in these last ten years had he known a genuinely carefree moment! Never! He said that he felt as if he were continually looking down the jaws of the monster.

> Letter to Eisenhower from the wife of Mark Mills of the Livermore Laboratory, following his death in a helicopter accident shortly before the HARDTACK series of nuclear tests in April 1958.[11]

Total War

If ever the world is blown to bits by some superbomb, there will be those who will watch the spectacle to the last minute, without fear, disinterestedly and with detachment.

J. Glenn Gray, *The Warriors* (1959).[12]

"I am sure that at the end of the world—in the last millisecond of the earth's existence—the last man will see what we have just seen."

George Kistiakowsky at Los Alamos, July 17, 1945, the day after the successful Trinity test of the first atomic bomb. (Asked if he still felt the same way during an interview in 1981, the ailing Kistiakowsky slammed his hand down on his desk and boomed, "YES!") [13]

Since that night [March 14, 1948] I had been daily steeped in alarming top secret intelligence. My mind was filled with a comparison between Hitler in the late 1930's and the encircling octopus of Communism in the late 1940's.

Robert Cutler, Eisenhower's first National Security Adviser.[14]

"The worst to be feared and the best to be expected can be simply stated. The *worst* is atomic war. The *best* would be this: a life of perpetual fear and tension; a burden of arms draining the wealth and labor of all peoples; a wasting of strength that defies the American system or the Soviet system or any system to achieve true abundance and happiness for the peoples of this earth."

Eisenhower, in April 1953. [15]

And if we could read aright the portent of this absence of guilt feelings in most modern soldiers, it would not be difficult to predict what is yet in store for us in the twentieth century.

J. Glenn Gray, *The Warriors* (1959). [16]

"My country had lost two wars in my young lifetime. The next time, I wanted to be on the winning side."

The German rocket scientist Wernher von Braun, explaining why he chose to surrender to the Americans. [17]

"I feel myself driven towards an end that I do not know."

Napoleon, on his way to Moscow in 1812. [18]

"It will run away with him, as it ran away with me."

Kaiser Wilhelm, speaking of Hitler to an English visitor to his place of exile in Doorn, the Netherlands, in the 1930's. [19]

141

Total War

Men have gained control over the forces of nature to such an extent that with their help they would have no difficulty in exterminating one another to the last man. They know this, and hence comes a large part of their current unrest, their unhappiness and their mood of anxiety.

Sigmund Freud, *Civilization and Its Discontents* (1929).[20]

Of the twenty or so civilizations known to modern Western historians, all except our own appear to be dead or moribund, and, when we diagnose each case, *in extremis* or *post mortem*, we invariably find that the cause of death has been either War or Class or some combination of the two. . . . We are thus confronted with a challenge that our predecessors never had to face: We have to abolish War and Class—and abolish them now—under pain, if we flinch or fail, of seeing them win a victory over man which, this time, would be conclusive and definitive.

Arnold Toynbee, *Civilization on Trial* 1948).[21]

The most ominous thing about these wars is that they were not isolated or unprecedented calamities. They were two wars in a series; and, when we envisage the whole series in a synoptic view, we discover that this is not only a series but also a progression. In our recent Western history war has been following war in an ascending order of intensity; and to-day it is already apparent that the War of 1939–45 was not the climax of this crescendo movement.

Arnold Toynbee, *War and Civilization* (1950).[22]

When I read the flood of scenarios . . . I ask myself in bewilderment: this war they are describing, *what is it about?* The defense of Western Europe? Access to the [Persian] Gulf? The protection of Japan?

The British historian Michael Howard, "On Fighting a Nuclear War" (1981).[23]

Qui desiderat pacem, praeparet bellum.
   [Who desires peace, let him prepare for war.]

Vegetius, *De Re Militari, Book iii, Prologue*, fourth century A.D.[24]

A wiser rule would be to make up your mind soberly what you want, peace or war, and then to get ready for what you want; for what we prepare for is what we shall get.

William Graham Sumner, "War" (1903).[25]

NOTES

1 THE ATOM

1. Ronald Clark, *Greatest Power on Earth*, p. 16, quoting *Blackwood's Magazine,* June 1914, p. 864.
2. Ronald Clark, *Greatest Power on Earth*, p. 23, quoting J. H. Thirring, *The Ideas of Einstein's Theory* (London: Methuen, 1921), p. 92.
3. Ronald Clark, *Einstein,* p. 101, quoting Philipp Frank, *Einstein: His Life and Times,* tr. George Rosen (New York: Alfred A. Knopf 1947), p. 211.
4. C. P. Snow, "A New Means of Destruction?" *Discovery,* London, September 1939, quoted in O'Keefe, pp. 31–32.
5. Blumberg and Owens, p. 120.
6. Ronald Clark, *Greatest Power on Earth,* p. 149.
7. *Ibid.* p. 176.
8. *Ibid.* p. 175.
9. *Ibid.* p. 177.
10. Hewlett and Anderson, p. 327, quoting the American

copy, papers of Franklin D. Roosevelt, Franklin D. Roosevelt Library, Hyde Park, NY.

11. Roberts, p. 9.

12. Sherwin, p. 138, quoting Henry L. Stimson Diaries, Sterling Memorial Library, Yale University, New Haven, CT.

13. James Byrnes, *All in One Lifetime* (New York: Harper & Brothers, 1958), p. 283, quoted in Ronald Clark, *Greatest Power on Earth,* p. 193.

14. Alice Kimball Smith, p. 480.

15. Herbert Feis, *Japan Subdued: The Atomic Bomb and the End of the War in the Pacific* (Princeton, NJ: Princeton University Press, 1961), p. 43, quoted in Alice Kimball Smith, p. 50.

16. Davis, p. 309.

17. Blumberg and Owens, p. 266.

18. Truman, p. 56.

19. Sherwin, p. 224, quoting Henry L. Stimson Diaries, Sterling Memorial Library, Yale University, New Haven, CT.

20. Francis Williams, p. 74.

21. Georgi K. Zhukov, *The Memoirs of Marshal Zhukov* (London: Jonathan Cape, 1971), p. 675, quoted in Prados, p. 16.

22. *The New York Times,* September 9, 1945, p. 1.

23. Thomas and Witts, p. 265.

24. Hachiya, pp. 13–14.

25. *Ibid.* pp. 6–7.

26. *Ibid.* p. 157.

27. Hahn, p. 170.

28. Irving, p. 17.

29. *Public Papers of the Presidents: Harry S. Truman, 1945,* August 9, p. 97.
30. *Time,* August 20, 1945, p. 36.
31. Herken, *Counsels of War,* p. 5.
32. Alice Kimball Smith, p. 77.
33. *Time,* October 29, 1945, p. 30.
34. Stern, p. 90.
35. *Ibid.*
36. Ulam, p. 224.
37. William L. Laurence, *The New York Times,* August 4, 1946, p. 3.
38. David Alan Rosenberg, "American Atomic Strategy and the Hydrogen Bomb Decision," *The Journal of American History,* Vol. 66, No. 1, June 1979, p. 67, quoting Papers of the Chief of Staff of the Air Force.
39. Ulam, p. 210.
40. *The Forrestal Diaries,* ed. Walter Mills (New York: Viking, 1959), p. 538, quoted in Beard, p. 30.
41. Herken, *Winning Weapon,* p. 30.
42. Ronald Clark, *Greatest Power on Earth,* p. 257.
43. *In the Matter of J. Robert Oppenheimer,* p. 467.
44. Ulam, p. 209.
45. Davis, p. 326.
46. Ronald Clark, *Greatet Power on Earth,* p. 258, quoting David E. Lilienthal, *Change, Hope and the Bomb* (Princeton, NJ: Princeton University Press, 1963), p. 143.
47. Dyson, p. 194.
48. Ulam, p. 217.
49. Ulam, p. 222.
50. Divine, p. 12, quoting Hagerty diary, Dwight D. Eisenhower Library, Abilene, KS.

51. Gerard Smith, p. 7.
52. Khrushchev, *Last Testament,* p. 52.
53. Seaborg, p. 32.
54. Ronald Clark, *Einstein,* p. 554.

2 THE MISSILES

1. Markov, p. 9.
2. Livingstone, p. 182.
3. Gavin, p. 144.
4. Montesquieu, Letter 105, p. 192.
5. Johnson, p. 345.
6. Williams and Epstein, p. 24.
7. *Ibid.* p. 20.
8. von Kármán, p. 243.
9. *Ibid.*
10. Koppes, p. 9.
11. Schaffer, p. 16, quoting Thomas M. Coffey, *Hap* (New York: Viking, 1982), p. 353.
12. Walter R. Dornberger, "The German V-2," Emme, p. 33.
13. Beard, p. 69.
14. Brown, p. 32.
15. Beard, p. 39.
16. Fuller, *Armament and History* (London: Eyre & Spottiswoode, 1946), p. 194.
17. G. A. Tokaty, "Soviet Rocket Technology," *Technology and Culture,* Vol. IV, No. 4 (Fall 1963), p. 523, quoted in Lasby, p. 297.
18. G. A. Tokaty, "Soviet Rocket Technology," Emme, p. 281.

19. Dwight D. Eisenhower Library, Abilene, KS.
20. *Khrushchev Remembers,* p. 405.
21. Koppes, p. 83.
22. Gavin, p. 260.
23. Averell Harriman, "My alarming interview with Khrushchev," *Life,* July 13, 1959, p. 33.
24. Baar and Howard, p. 43.
25. Baar and Howard, p. 226.
26. *Khrushchev Remembers,* p. 493.
27. Mandelbaum, p. 139.
28. Roberts, p. 5.
29. Bohlen, p. 496.
30. "The Atomic Age," *Time,* July 29, 1985, p. 51.
31. Charles Draper letter, *The New York Times,* September 20, 1981, Section 4, p. 20.
32. Heikal, p. 154.
33. Brzezinski, p. 304.

3   HOW MUCH IS ENOUGH?

1. Tuchman, *Proud Tower,* p. 380.
2. Churchill, *World Crisis,* Vol. I, p. 33.
3. Liddell Hart, *Why Don't We Learn from History?,* p. 67.
4. Fuller, *The Conduct of War,* p. 241.
5. Zuckerman, p. 142.
6. Coffey, p. 42.
7. *The Army Air Forces in World War II,* ed. Wesley Frank Craven and James Lee Cate (Chicago: The University of Chicago Press, 1951), Vol. III, p. 733, quoted in Hastings, p. 339.
8. Schaffer, p. xiii.

9. Herken, *Winning Weapon,* p. 262.
10. Schaffer, p. 197, quoting House Committee on Armed Services, "Unification Hearings," 81st Congress, 1st session.
11. Samuel P. Huntington, *The Common Defense: Strategic Programs in National Politics* (New York: Columbia University Press, 1961), p. 369, quoted in Beard, p. 39.
12. *The New York Times,* October 12, 1949, p. 34, quoted in Stern, p. 147.
13. James R. Shepley and Clay Blair, Jr., *The Hydrogen Bomb* (Westport, CT: Greenwood Press, 1971), p. 81, quoted in Stern, p. 148.
14. Moss, p. 117.
15. Trewhitt, p. 72.
16. Power, p. 132.
17. Murphy, p. 391.
18. Blackstock, *Strategy of Subversion,* p. 31, quoting U.S. Department of State *Bulletin,* June 18, 1956, pp. 999–1004.
19. John L. Chapman, p. 177.
20. Kistiakowsky, p. 160.
21. Cockburn, p. 452.
22. Beschloss, p. 223.
23. Kistiakowsky, p. 293.
24. Trewhitt, p. 115.
25. Herken, *Counsels of War,* p. 168.
26. Enthoven and Smith, p. 197.
27. Trewhitt, p. 117.
28. Herken, *Counsels of War,* p. 206.
29. Gerard Smith, p. 25.
30. Cockburn, p. 15.
31. Trewhitt, p. 13.
32. *Ibid.* p. 17.

33. Freedman, p. 85, quoting Robert McNamara, "The Dynamics of Nuclear Strategy," *Department of State Bulletin* LVII, October 9, 1967, p. 445.
34. McGeorge Bundy, "To Cap the Volcano," *Foreign Affairs,* Vol. 48 No. 1 (October 1969), p. 10.
35. Kissinger, p. 84.
36. "War and Peace in the Nuclear Age," ed. Michael C. Janeway and Harry W. King, *Boston Globe,* October 17, 1982, p. 8.
37. *Ibid.*
38. Howe, p. 49.
39. Kissinger, p. 67.
40. *Department of State Bulletin,* July 29, 1974, p. 215, quoted in Prados, p. 239.
41. Nixon, p. 151.

4   THE RUSSIANS

1. Wallace, p. 274.
2. Blainey, p. 89.
3. Cooper, p. 2.
4. *Ibid.* p. 83.
5. Brzezinski, p. 541.
6. de Tocqueville, p. 412.
7. Custine, p. 330.
8. *Ibid.* p. 50.
9. *Ibid.* p. 76.
10. Creasy, p. 282.
11. Karl Marx, *New York Tribune,* December 30, 1853, quoted in Mastny, p. 3.
12. Lincoln, p. 149.

13. Holloway, p. 14.
14. Egmont Zechlin, "Cabinet versus Economic Warfare in Germany," Koch, p. 149.
15. Aron, p. 209.
16. Power, p. 36.
17. Boris Souvarine, *Stalin: A Critical Survey of Bolshevism* (New York: Longmans, Green, 1939), p. 485, quoted in Blackstock, p. 284.
18. Marie, Grand Duchess of Russia, p. 220.
19. Edmund Wilson, p. 258.
20. Liddell Hart, Memoirs, p. 291.
21. Churchill, *Complete Speeches,* Vol. VI, p. 6161.
22. *Khrushchev Remembers*, p. 170.
23. Zuckerman, p. 215.
24. Colville, p. 563.
25. Gavin, p. 87.
26. Ronald Clark, *Greatest Power on Earth,* p. 218, quoting Leo Szilard, "A Personal History of the Atomic Bomb," University of Chicago *Round Table,* September 25, 1949, pp. 14–15.
27. Sherwin, p. 212.
28. Roberts, p. 18.
29. Pringle and Spigelman, p. 63.
30. Murphy, p. 164.
31. Cutler, p. 250.
32. *New York Times,* October 10, 1948, p. 3, quoted in Kissinger, p. 63.
33. Holloway, p. 132.
34. *Khrushchev Remembers,* p. 368.
35. Salisbury, p. 430.

36. Bohlen, p. 356.
37. *In the Matter of J. Robert Oppenheimer,* p. 363.
38. Holloway, p. 146.
39. Khrushchev, *Last Testament,* p. 288.
40. Mićunović, p. 93.
41. Heikal, p. 128.
42. Bohlen, p. 437.
43. Mićunović, p. 432.
44. Khrushchev, *Last Testament,* p. 33.
45. Sejna, p. 49.
46. York, p. 122.
47. Bialer, p. 12.
48. Sejna, p. 120.
49. Powers, p. 86.
50. Cockburn, p. 235.
51. *Ibid.* p. 176.
52. Kissinger, p. 413.
53. Bialer, p. 260.
54. Hugh Sidey, "The Presidency," *Time,* October 20, 1986, p. 31.

5 . ARMS CONTROL

1. Morland, p. 231.
2. Gooch, p. 226.
3. Kenneth Boulding, *Beyond Economics: Essays on Society, Religion, and Ethics* (Ann Arbor, MI: University of Michigan Press, 1968), p. 105, quoted in Blainey, p. 31.
4. Howard, *War in European History,* p. 24.
5. Lincoln, p. 232.

6. Tuchman, *Proud Tower,* p. 287, quoting Joseph Hodges Choate, *The Two Hague Conferences* (Princeton, NJ: Princeton University Press, 1913), p. 40.
7. Temperley, p. 53.
8. Liddell Hart, *Memoirs,* p. 202.
9. Temperley, p. 120.
10. *Ibid.* p. 276.
11. Bohlen, p. 213.
12. *Robert Oppenheimer: Letters and Recollections,* p. 320.
13. Herken, *Winning Weapon,* p. 221.
14. *Ibid.* pp. vii, 178.
15. *Ibid.* p. 161.
16. Davis, p. 260.
17. Bernard M. Baruch, *The Public Years* (New York: Holt, Rinehart and Winston, 1960), p. 369, quoted in Ronald Clark, *The Greatest Power on Earth,* p. 247.
18. *A Tribute to Niels Bohr on the Hundredth Anniversary of His Birth,* ed. M. Jacob (Geneva: CERN, 1985), p. 11.
19. Stern, p. 147.
20. Herken, *Winning Weapon,* p. 328.
21. J. Robert Oppenheimer, "Atomic Weapons and American Policy," *Foreign Affairs,* Vol. 31, No. 4 (July 1953), p. 525.
22. Kissinger, p. 522.
23. Aron, p. 200.
24. Kistiakowsky, p. 181.
25. Gerard Smith, p. 36.
26. Seaborg, p. 6, quoting Philip Noel-Baker, *The Arms Race,* quoted in Bernford G. Bechoefer, *Postwar Negotiations for Arms Control* (Washington, DC: The Brookings Institution, 1961).

27. Khrushchev, *Last Testament,* p. 411.
28. Divine, p. 192.
29. Blumberg and Owens, p. 296.
30. Davis, p. 329.
31. Trewhitt, p. 21.
32. Roberts, p. 64.
33. Murphy, p. 435.
34. Roberts, p. 6.
35. Robert Kleiman, "Ban Space-Weapons Tests," *The New York Times,* December 10, 1984, p. 23.
36. Prados, p. 210.
37. Gerard Smith, p. 24.
38. *Ibid.* p. 263.
39. Freedman, p. 168.
40. Cohen, p. 107.
41. *The New York Times,* December 21, 1978, p. A3.
42. Gerard Smith, p. 13.
43. Scheer, p. 90.
44. Talbott, p. 85.
45. *Ibid.* p. 348.

6 MONEY, SECRETS, AND SCIENTISTS

1. Sir George Clark, p. 134.
2. Howard, *War in European History,* p. 38.
3. Lincoln, p. 231.
4. Howard, *War in European History,* p. 48.
5. Ellis, p. 27.
6. Bloch, p. xii.
7. Kevles, p. 156.
8. Khrushchev, *Last Testament,* p. 411.

9. de Gaulle, p. 232.
10. Trewhitt, p. 122.
11. *Seminar on C³I* (1981), p. 135.
12. Gerard Smith, p. 222.
13. Bingham, p. 405.
14. *Ibid.* p. 418.
15. Marc Bloch, *Strange Defeat* (Oxford University Press, 1949), pp. 83–84, quoted in MacIsaac, p. 214.
16. *Seminar on C³I* (1981), p. 126.
17. Blackstock, *Strategy of Subversion,* p. 106.
18. Cockburn, p. 16, quoting PBS-TV *World,* "The Red Army," copyright 1981 WGBH Educational Foundation, Boston, and Granada Television International Ltd., London.
19. *Seminar on C³I* (1981), p. 198.
20. Ransom, p. 169, quoting *Congressional Record,* April 9, 1956, p. 5292.
21. de Gramont, p. 16.
22. Rositzke, p. 37.
23. Hood, p. 151.
24. Felix, p. 30.
25. Salinger, p. 155.
26. Cline, p. 58.
27. *Seminar on C³I* (1981), p. 209.
28. Corson and Crowley, p. 133, quoting S. Vladimir Krulich, *Los Angeles Times,* March 6, 1977.
29. Bradley Smith, p. 391.
30. Fuller, *Armament and History,* p. 100.
31. Kevles, p. 141, quoting T. F. Crane, "Scholars and the War—Then and Now," *The Nation,* August 3, 1916, p. 107.

32. *The New York Times,* May 10, 1931, p. 3, quoted in Kevles, p. 28.
33. Gissing, p. 241.
34. Kevles, p. 113.
35. *Ibid.* p. 155, quoting *Report of the President of Columbia University, 1919,* p. 147.
36. *Ibid.* p. 279.
37. Goudsmit, p. 112.
38. Ulam, p. 58.
39. Blumberg and Owens, p. 296.
40. Herken, *Counsels of War,* p. 235.

7   THE NATURE OF WAR

1. Leakey, p. 169.
2. Wellard, p. 101.
3. Hippocrates, pp. 115–117.
4. Sun Tzu, pp. 77–79.
5. Cooper, p. 16.
6. Toynbee, *War and Civilization,* p. 108.
7. Gaster, p. 313.
8. Garlan, p. 172, quoting Vegetius, I, 13.
9. Fuller, *Armament and History,* p. 47, quoting Vegetius i:I.
10. Livingstone, p. 181.
11. Tacitus, p. 695 and *Oxford Dictionary of Quotations,* p. 531.
12. Villari, p. 110.
13. Chambers, p. 6.
14. Seward, p. 101, quoting Jean de Venette, *The Chronicle of Jean de Venette,* tr. J. Birdsall (New York, Columbia University Press, 1953).

15. Seward, p. 15.
16. Sykes, p. 175.
17. *Oxford Dictionary of Quotations,* p. 563, quoting W. Coxe, *Memoirs of Sir Robert Walpole* (1798).
18. Piozzi, p. 88.
19. Franklin, p. 11.
20. Howard, *War in European History,* p. 81.
21. Parkinson, p. 55.
22. Clausewitz, p. 63.
23. *Ibid.* p. 260.
24. *Ibid.* p. 69.
25. de Tocqueville, p. 657.
26. Koch, p. 340.
27. Barzini, p. 64.
28. Cave Brown, p. 354.
29. Sherman, p. 14.
30. *Ibid.* p. 13.
31. Andrew Wilson, p. 132.
32. Sherman, p. 7.
33. Creasy, p. 340, quoting Goethe, "Campaign in France in 1792," tr. Farie, p. 77.
34. Gerard Smith, p. 67.
35. Macmillan, p. 352.
36. Sherman, p. 59.
37. Walters, p. 114.
38. Sherman, p. 105.
39. Forbes, p. 241.
40. Toland, p. 450.
41. Crankshaw, p. 133.

42. McFeely, p. 78, quoting O. O. Howard Paper, Bowdoin College Library, Brunswick, ME.
43. Parkinson, p. 337.
44. Gray, p. 31.
45. Shepherd, p. 75.
46. Lady Gwendolen Cecil, *Life of Robert, Marquis of Salisbury,* Vol. II: 1868–1880 (London, 1921), p. 153, quoted in *Oxford Dictionary of Quotations,* p. 413.
47. Blainey, p. 213.
48. Toynbee, *War and Civilization,* p. 16.
49. McFeely, p. 30.
50. "Editorial," *Life,* October 10, 1949, p. 38.
51. Sergei Witte, *The Memoirs of Count Witte* (London: Yarmolinsky, 1921), p. 250, quoted in Blainey, p. 76.
52. Taylor, *Politics in Wartime,* p. 97.
53. Barzini, p. 59.
54. Howard, *The Causes of Wars,* p. 169.
55. Liddell Hart, *Why Don't We Learn from History?,* p. 62.
56. Freud, *Civilisation, War and Death,* p. 93.
57. Ferrero, p. 63.
58. Jukić, p. 57.
59. Djilas, p. 114.
60. Toynbee, *War and Civilization,* p. viii.
61. Mills, p. 58.
62. Power, p. 229.
63. Raymond Aron, *Peace and War: A Theory of International Relations,* tr. Richard Howard and Annette Baker Fox (New York: Doubleday, 1966), p. 7, quoted in Howard, *The Causes of Wars,* p. 7.
64. Liddell Hart, *Why Don't We Learn from History?* p. 58.

65. Morris, p. 75.
66. Lansdale, p. 71.
67. *Seminar on C³I* (1981), p. 134.
68. Howard, *The Causes of Wars,* p. 21.
69. *Ibid.* p. 203.

8 THE CENTURY OF TOTAL WAR

1. Creasy, p. 346.
2. Sherman, p. 9.
3. Blainey, p. 136.
4. Forbes, p. 248.
5. Tuchman, *Guns of August,* p. 71.
6. Koch, p. 346.
7. Phillips, p. 526.
8. Sumner, p. 29.
9. James Joll, "The Unspoken Assumptions," Koch, p. 317.
10. Churchill, *Complete Speeches,* Vol. II, p. 1085.
11. Tuchman, *Guns of August,* p. 49, quoting Major-General Sir Charles E. Calwell, *Field Marshal Sir Henry Wilson: His Life and Diaries* (New York: Scribner's, 1927), Vol. 1, p. 78.
12. Lincoln, p. 416.
13. Taylor, *How Wars Begin,* p. 100.
14. Karl-Heinz Jannsen, "Gerhard Ritter: A Patriotic Historian's Justification," Koch, p. 261.
15. James, p. 713.
16. James Joll, "The Unspoken Assumptions," Koch, p. 325.
17. Liddell Hart, *Memoirs,* p. 21.
18. Grigorenko, p. 5.

19. Vivien Raynor, *The New York Times,* May 24, 1985, p. C36.
20. Guy Chapman, p. 218.
21. Ellis, p. 50.
22. Toland, p. 150.
23. Rudolf Binding, *A Fatalist at War,* tr. Ian F. D. Morrow (London: George Allen & Unwin, 1929), p. 216, quoted in Toland, p. 119.
24. Toland, p. 578.
25. Blainey, p. 6.
26. Hastings, p. 43.
27. Blumberg and Owens, p. 50.
28. Lippmann, p. 90.
29. *War in Our Time,* ed. with an introduction by Hans Speier and Alfred Kähler (New York: Norton, 1939), p. 13, quoted in Fuller, *Armament and History,* p. 165.
30. Hastings, p. 94.
31. Horstmann, p. 50.
32. Hastings, p. 107.
33. Churchill, *Complete Speeches,* Vol. VII, p. 7800.
34. *Foreign Relations of the United States,* 1950, Vol. 1, p. 143.
35. Herken, *Counsels of War,* p. 49.
36. Gavin, p. 124.
37. Divine, p. 13.
38. Beschloss, p. 84, quoting transcript of tape-recorded conversation, Dwight D. Eisenhower Presidential Papers, Dwight D. Eisenhower Library, Abilene, KS.
39. *Meeting the Threat of Surprise Attack,* Report of the Technological Capabilities Panel of the Science Advisory Committee, February 14, 1955.
40. *Ibid.*

41. Trewhitt, p. 75.
42. Herken, *Counsels of War*, p. 116.
43. Heikal, p. 97.
44. Nixon, p. 178.
45. Beschloss, p. 117.
46. Seaborg, p. 127.
47. Trewhitt, p. 184. McNaughton was head of Internal Security Affairs at the Pentagon.
48. "The Atomic Age," *Time*, July 29, 1985, p. 50.
49. Kissinger, p. 391.
50. Morris, p. 240.
51. Scheer, p. 5.
52. *Seminar on C^3I* (1980), p. 3.

9 FOREBODINGS

1. Johnson, p. 379.
2. Brodie, *Strategy in the Missile Age,* p. 6.
3. Fuller, *Armament and History,* p. xi.
4. Seldes, p. 41.
5. Parkinson, p. 346.
6. Brodie, *War and Politics,* pp. 27–28.
7. Tuchman, *Guns of August,* p. 122, quoting Grey, Viscount of Fallodon, *Twenty-five Years: 1892–1916,* Vol. II (London: Hodder & Stoughton, 1925), p. 20.
8. Taylor, *Politics in Wartime,* p. 14.
9. Douhet, p. 181.
10. Hughes, p. 203.
11. Divine, p. 205.
12. Gray, p. 36.
13. Laurence, p. 11.

14. Cutler, p. 249.
15. *Ibid.* p. 338.
16. Gray, p. 168.
17. Lang, p. 186.
18. Fuller, *The Conduct of War,* p. 43.
19. James Joll, "The Unspoken Assumptions," Koch, p. 308.
20. Freud, *Civilization and Its Discontents,* p. 92.
21. Toynbee, *Civilization on Trial,* pp. 23–25.
22. Toynbee, *War and Civilization,* p. 4.
23. Michael Howard, "On Fighting a Nuclear War," *International Security,* Spring 1981, p. 9.
24. *Home Book of Quotations,* p. 1599.
25. Sumner, p. 40.

BIBLIOGRAPHY

Aron, Raymond. *The Century of Total War.* New York: Doubleday, 1954.

Baar, James, and William E. Howard. *Polaris!* New York: Harcourt Brace, 1960.

Barzini, Luigi. *The Europeans.* New York: Simon and Schuster, 1983.

Beard, Edmund. *Developing the ICBM: A Study in Bureaucratic Politics.* New York: Columbia University Press, 1976.

Beschloss, Michael R. Mayday: *Eisenhower, Khrushchev and the U-2 Affair.* New York: Harper & Row, 1986.

Bialer, Seweryn. *Stalin's Successors: Leadership, Stability, and Change in the Soviet Union.* New York: Cambridge University Press, 1980.

Bingham, Colin. *Men and Affairs.* New York: Funk & Wagnalls, 1967.

Blackstock, Paul W. *The Secret Road to World War II: Soviet versus Western Intelligence.* Chicago: Quadrangle Books, 1969.

_____. *The Strategy of Subversion: Manipulating the Politics of Other Nations.* Chicago: Quadrangle Books, 1964.

Blainey, Geoffrey. *The Causes of War.* New York: Free Press, 1973.

Bloch, Ivan. *The Future of War.* Boston: Ginn & Co., 1902.

Blumberg, Stanley A., and Gwinn Owens. *Energy and Conflict: The Life and Times of Edward Teller.* New York: G. P. Putnam's Sons, 1976.

Bohlen, Charles E. *Witness to History, 1929–1969.* New York: W. W. Norton, 1973.

Brodie, Bernard. *Strategy in the Missile Age.* Princeton, NJ: Princeton University Press, 1959.

Bibliography

———. *War and Politics*. New York: Macmillan, 1974.

Brown, Harrison. *Must Destruction Be Our Destiny? A Scientist Speaks as a Citizen*. New York: Simon and Schuster, 1946.

Brzezinski, Zbigniew. *Power and Principle: Memoirs of the National Security Adviser 1977–1981*. New York: Farrar Strauss Giroux, 1983.

Cave Brown, Anthony. *The Last Hero: Wild Bill Donovan*. New York: Times Books, 1982.

Chambers, James. *The Devil's Horsemen: The Mongol Invasion of Europe*. New York: Atheneum, 1979.

Chapman, Guy. *A Passionate Prodigality: Fragments of Autobiography* (1933). First American edition. New York: Holt, Rinehart and Winston, 1965.

Chapman, John L. *Atlas: The Story of a Missile*. New York: Harper and Brothers, 1960.

Churchill, Winston S. *Winston S. Churchill: His Complete Speeches: 1897–1963;* ed. Robert Rhodes James. New York: Chelsea House/R. R. Bowker, 1974.

———. *The World Crisis*. New York: Charles Scribner's Sons, 1923; renewal © 1951, Winston S. Churchill.

Clark, Sir George. *War and Society in the Seventeenth Century*. New York: Cambridge University Press, 1958.

Clark, Ronald W. *Einstein: The Life and Times*. New York: World Publishing, 1971.

———. *The Greatest Power on Earth: The International Race for Nuclear Supremacy*. New York: Harper & Row, 1980.

Clausewitz, Carl von. *On War,* ed. and trans. Michael Howard and Peter Paret. Princeton,NJ: Princeton University Press, 1976.

Cline, Ray S. *Secrets, Spies, and Scholars: Blueprint of the Essential CIA*. Washington, DC: Acropolis Books, 1976.

Cockburn, Andrew. *The Threat: Inside the Soviet Military Machine*. New York: Random House, 1983.

Bibliography

Coffey, Thomas M. *Iron Eagle: The Turbulent Life of General Curtis LeMay.* New York: Crown, 1986.

Cohen, Sam. *The Truth About the Neutron Bomb: The Inventor of the Bomb Speaks Out.* New York: William Morrow, 1983.

Colville, John Rupert, Sir. *The Fringes of Power: Downing Street Diaries. 1939–1955.* London: Hodder and Stoughton, 1985.

Cooper, Leonard. *Many Roads to Moscow: Three Historic Invasions.* New York: Coward-McCann, 1968.

Corson, William R., and Robert T. Crowley. *The New KGB: Engine of Soviet Power.* New York: William Morrow, 1985.

Crankshaw, Edward. *Bismarck.* New York: Viking Press, 1981.

Creasy, E. S. *The Fifteen Decisive Battles of the World from Marathon to Waterloo.* New York: Harper & Brothers, 1851.

Custine, Marquis de. *The Journals of the Marquis de Custine: Journey for Our Time,* ed. and trans. Phyllis Penn Kohler. New York: Pellegrini & Cudahy, 1951.

Cutler, Robert. *No Time for Rest.* Boston: Little, Brown, 1966.

Davis, Nuel Pharr. *Lawrence and Oppenheimer.* New York: Simon and Schuster, 1968.

Divine, Robert A. *Blowing on the Wind: The Nuclear Test Ban Debate: 1954–1960.* New York: Oxford University Press, 1978.

Djilas, Milovan. *Conversations with Stalin,* trans. Michael B. Petrovich. New York: Harcourt, Brace & World, 1962.

Douhet, Giulio. *The Command of the Air,* trans. Dino Ferrari. New York: Coward-McCann, 1942.

Dyson, Freeman J. *Disturbing the Universe.* New York: Harper & Row, 1979.

Ellis, Havelock. *Essays in War-time: Further Studies in the Task of Social Hygiene.* Boston: Houghton Mifflin, 1917.

167

Bibliography

Emme, Eugene M., ed. *The History of Rocket Technology.* Detroit: Wayne State University Press, 1964.

Enthoven, Alain C., and Wayne K. Smith. *How Much Is Enough? Shaping the Defense Program, 1961–1969.* New York: Harper & Row, 1971.

Felix, Christopher [pseud.]. *The Spy and His Masters: A Short Course in the Secret War.* London: Secker & Warburg, 1963.

Ferrero, Guglielmo. *Peace and War,* trans. Bertha Pritchard. London: Macmillan, 1933.

Foreign Relations of the United States. Washington, DC: U.S. Government Printing Office, 1953.

Forbes, Archibald. *Memories and Studies of War and Peace.* London: Cassell, 1895.

Franklin, Benjamin. *The Works of Benjamin Franklin: with notes and a life of the author by Jared Sparks,* Vol. X. Chicago: Townsend Mac Coun, 1882.

Freedman, Lawrence. *U.S. Intelligence and the Soviet Strategic Threat.* Boulder, CO: Westview Press, 1977.

Freud, Sigmund. *Civilization and Its Discontents,* tr. James Strachey. New York: W. W. Norton, 1961.

_____. *Civilisation, War and Death,* ed John Rickman. London: The Hogarth Press and the Institute of Psychoanalysis, 1968.

Fuller, Major General J.F.C. *Armament and History: A Study of the Influence of Armament on History from the Dawn of Classical Warfare to the Second World War.* New York: Charles Scribner's Sons, 1945.

_____. *The Conduct of War: 1789–1961: A Study of the Impact of the French, Industrial, and Russian Revolutions on War and Its Conduct.* New Brunswick, NJ: Rutgers University Press, 1961.

Garlan, Yvon. *War in the Ancient World: A Social History,* trans. Janet Lloyd. New York: W. W. Norton, 1975.

168

Bibliography

Gaster, Theodor H., ed. *The Dead Sea Scriptures in English Translation.* New York: Doubleday, 1956.

Gaulle, Charles de. *Memoirs of Hope: Renewal and Endeavor,* tr. Terence Kilmartin. New York: Simon and Schuster, 1979.

Gavin, Lt. General James M. *War and Peace in the Space Age.* New York: Harper & Brothers, 1958.

Gissing, George Robert. *The Private Papers of Henry Ryecroft: Author's Only Authorized American Edition.* New York: E. P. Dutton, 1927.

Gooch, G. P. *Studies in Diplomacy and Statecraft.* London: Longmans, Green, 1942.

Goudsmit, Samuel A. *Alsos.* New York: Henry Schuman, 1947.

de Gramont, Sanche (Ted Morgan). *The Secret War: The Story of International Espionage Since World War II.* New York: G. P. Putnam's Sons, 1962.

Gray, J. Glenn, *The Warriors: Reflections on Men in Battle.* New York: Harcourt Brace, 1959.

Grigorenko, Petro G. *Memoirs,* trans. Thomas P. Whitney, New York: W. W. Norton, 1982.

Hachiya, M.D., Michihiko. *Hiroshima Diary: The Journal of a Japanese Physician: August 6–September 30, 1945,* trans. Warner Wells, M.D. Chapel Hill, NC: University of North Carolina Press, 1955.

Hahn, Otto. *My Life: The Autobiography of a Scientist,* trans. Ernest Kaiser and Eithne Wilkins. New York: Herder and Herder, 1970.

Hastings, Max. *Bomber Command.* New York: Dial Press/James Wade, 1977.

Heikal, Mohamed. *The Sphinx and the Commissar: The Rise and Fall of Soviet Influence in the Middle East.* New York: Harper & Row, 1978.

Herken, Gregg. *Counsels of War.* New York: Alfred A. Knopf, 1985.

_____. *The Winning Weapon: The Atomic Bomb in the Cold War.* New York: Alfred A. Knopf, 1980.

169

Bibliography

Hewlett, Richard G., and Oscar E. Anderson, Jr. *The New World: 1939/1946, Vol. I: A History of the United States Atomic Energy Commission.* University Park, PA: Pennsylvania State University Press, 1962.

Hippocrates, "Airs Waters Places," *Hippocrates,* Vol. I, trans. W. H. S. Jones, The Loeb Classical Library. London: William Heinemann, 1923.

Holloway, David. *The Soviet Union and the Arms Race.* New Haven, CT: Yale University Press, 1983.

The Home Book of Quotations: Classical and Modern, ed. Burton Stevenson. New York: Dodd, Mead, 1934.

Hood, William. *Mole.* New York: W. W. Norton, 1982.

Horstmann, Lali. *We Chose to Stay.* Boston: Houghton Mifflin, 1954.

Howard, Michael. *The Causes of Wars and Other Essays,* 2nd ed. London: Temple Smith, 1983.

————. *War in European History.* New York: Oxford University Press, 1976.

Howe, Russell Warren. *Weapons, the International Game of Arms, Money, and Diplomacy.* New York: Doubleday, 1980.

Hughes, Emmet John. *The Ordeal of Power: A Political Memoir of the Eisenhower Years,* paperback ed. New York: Atheneum, 1975.

Irving, David. *The German Atomic Bomb: The History of Nuclear Research in Nazi Germany.* New York: Simon and Schuster, 1967.

James, Henry. *Henry James Letters,* Vol. IV, 1895–1916, ed. Leon Edel. Cambridge, MA: Belknap Press of Harvard University Press, 1984.

Johnson, Samuel, "The History of Rasselas, Prince of Abyssinia" (1759), *The Oxford Authors: Samuel Johnson,* ed. Donald Greene. New York: Oxford University Press, 1984.

Jukić, Ilija. *The Fall of Yugoslavia.* New York: Harcourt Brace Jovanovich, 1974.

Bibliography

von Kármán, Theodor, with Lee Edson. *The Wind and Beyond: Theodore von Kármán: Pioneer in Aviation and Pathfinder in Space.* Boston: Little, Brown, 1967.

Kevles, Daniel J. *The Physicists: The History of a Scientific Community in Modern America.* New York: Alfred A. Knopf, 1977.

Khrushchev, Nikita. *Khrushchev Remembers,* ed. and trans. Strobe Talbott. Boston: Little, Brown. 1970.

———. *Khrushchev Remembers: The Last Testament,* ed. and trans. Strobe Talbott. Boston: Little, Brown, 1974.

Kissinger, Henry. *White House Years.* Boston: Little, Brown, 1979.

Kistiakowsky, George B. *A. Scientist at the White House: The Private Diary of President Eisenhower's Special Assistant for Science and Technology.* Cambridge, MA: Harvard University Press, 1976.

Koch, H. W., ed. *The Origins of the First World War: Great Power Rivalry and German War Aims.* New York: Taplinger, 1972.

Koppes, Clayton R. *JPL and the American Space Program: A History of the Jet Propulsion Laboratory.* New Haven, CT: Yale University Press, 1982.

Lang, Daniel. *From Hiroshima to the Moon: Chronicles of Life in the Atomic Age.* New York: Simon and Schuster, 1959.

Lansdale, Edward Geary, Major General, U.S.A.F. (Ret.). *In the Midst of Wars: An American's Mission to Southeast Asia.* New York: Harper & Row, 1972.

Lasby, Clarence G. *Project Paperclip: German Scientists and the Cold War.* New York: Atheneum, 1971.

Laurence, William L. *Dawn Over Zero: The Story of the Atomic Bomb.* New York: Alfred A. Knopf, 1946.

Leakey, Richard E. *The Making of Mankind.* New York: E. P. Dutton, 1981.

Liddell Hart, B. H. *The Liddell Hart Memoirs: 1895–1938,* Vol. 1. New York: G. P. Putnam's Sons, 1965.

Bibliography

_____. *Why Don't We Learn from History?* New York: Hawthorn Books, 1971.

Lincoln, W. Bruce. *In War's Dark Shadow: The Russians Before the Great War.* New York: Dial Press, 1983.

Lippmann, Walter. *The Good Society.* New York: Grossett's Universal Library, 1937.

Livingstone, R. W. *The Mission of Greece: Some Greek Views of Life in the Roman World.* London: Oxford University Press, 1928.

McFeely, William S. *Grant: A Biography.* New York: W. W. Norton, 1981.

MacIsaac, David. *Strategic Bombing in World War Two: The Story of The United States Strategic Bombing Survey.* New York: Garland Publishing, 1976.

Macmillan, Harold. *The Blast of War: 1939–1945.* New York: Harper & Row, 1967.

Mandelbaum, Michael. *The Nuclear Question: The United States and Nuclear Weapons, 1946–1976.* New York: Cambridge University Press, 1979.

Marie, Grand Duchess of Russia. *Education of a Princess,* trans. under editorial supervision of Russell Lord. New York: Viking Press, 1930.

Markov, M. A. *Science and the Responsibility of Scientists.* Moscow: Nauka Publishers, 1981.

Mastny, Vojtech. *Russia's Road to the Cold War: Diplomacy, Warfare, and the Politics of Communism, 1941–1945.* New York: Columbia University Press, 1979.

Mićunović, Veljko. *Moscow Diary,* trans. David Floyd. New York: Doubleday, 1980.

Mills, C. Wright. *The Causes of World War Three.* New York: Simon and Schuster, 1958.

Bibliography

Montesquieu. *Persian Letters* (1721), trans. C. J. Betts. Baltimore: Penguin, 1973.

Morland, Howard. *The Secret That Exploded.* New York: Random House, 1981.

Morris, Roger. *Uncertain Greatness: Henry Kissinger and American Foreign Policy.* New York: Harper & Row, 1977.

Moss, Norman. *Men Who Play God: The Story of the H-Bomb and How the World Came to Live with It.* New York: Harper & Row, 1968.

Murphy, Robert. *Diplomat Among Warriors.* New York: Doubleday, 1964.

Nixon, Richard. *The Real War.* New York: Warner Books, 1980.

O'Keefe, Bernard J. *Nuclear Hostages.* Boston: Houghton Mifflin, 1983.

In the Matter of J. Robert Oppenheimer: Transcript of Hearing Before Personnel Security Board. Washington, DC: U.S. Government Printing Office, 1954.

The Oxford Dictionary of Quotations, 3rd ed. New York: Oxford University Press, 1979.

Parkinson, Roger. *Clausewitz: A Biography.* London: Wayland Publishers, 1970.

Phillips, W. Alison. *Modern Europe: 1815–1899.* London: Rivingtons, 1901.

Piozzi, Hesther Lynch. *Anecdotes of the Late Samuel Johnson, LL.D., during the Last Twenty Years of His Life,* ed. Arthur Sherbo. London: Oxford University Press, 1974.

Power, General Thomas S., USAF (Ret.), with Albert A. Arnhym. *Design for Survival.* New York: Coward-McCann, 1964, 1965.

Powers, Thomas. *Thinking About the Next War.* New York: Alfred A. Knopf, 1982.

Prados, John. *The Soviet Estimate: U.S. Intelligence Analysis & Russian Military Strength.* New York: Dial Press, 1982.

Bibliography

Pringle, Peter, and James Spigelman. *The Nuclear Barons*. New York: Holt, Rinehart and Winston, 1981.

Ransom, Harry Howe. *The Intelligence Establishment*. Cambridge, MA: Harvard University Press, 1970.

Roberts, Chalmers M. *The Nuclear Years: The Arms Race and Arms Control, 1945–70*. New York: Praeger, 1970.

Rositzke, Harry August. *The CIA's Secret Operations: Espionage, Counter-espionage, and Covert Action*. New York: Readers Digest Press, 1977.

Salinger, Pierre. *With Kennedy*. New York: Doubleday, 1966.

Salisbury, Harrison E. *A Journey for Our Times: A Memoir*. New York: Harper & Row, 1983.

Schaffer, Ronald. *Wings of Judgment: American Bombing in World War II*. New York: Oxford University Press, 1985.

Scheer, Robert. *With Enough Shovels: Reagan, Bush and Nuclear War*. New York: Random House, 1982.

Seaborg, Glenn T., with the assistance of Benjamin S. Loeb. *Kennedy, Khrushchev and the Test Ban*. Berkeley, CA: University of California Press, 1961.

Sejna, Jan. *We Will Bury You*. London: Sidgwick & Jackson, 1982.

Seldes, George, compiler. *The Great Quotations*. New York: Lyle Stuart, 1960.

Seminar on Command, Control, Communications and Intelligence: Spring 1980, 1981. Cambridge, MA: Program on Information Resources Policy, Harvard University, Center for Information Policy Research, 1981. [*Seminar on C^3I* (1980) or (1981) in Notes.]

Seward, Desmond. *The Hundred Years War: The English in France, 1337–1453*. New York: Atheneum, 1978.

Shepherd, William G. *Confessions of a War Correspondent*. New York: Harper & Brothers, 1917.

Sherman, General William T. *"War Is Hell!"* from *Memoirs of General William T. Sherman by Himself,* D. Appleton and Company, New York (1875), ed. Mills Lane. Savannah, GA: Beehive Press, 1974.

Sherwin, Martin J. *A World Destroyed: The Atomic Bomb and the Grand Alliance.* New York: Alfred A. Knopf, 1975.

Smith, Alice Kimball. *A Peril and a Hope: The Scientists' Movement in America: 1945–1947.* Chicago: University of Chicago Press, 1965.

Smith, Bradley F. *The Shadow Warriors: O.S.S. and the Origins of the C.I.A.* New York: Basic Books, 1983.

Smith, Gerard. *Doubletalk: The Story of the First Strategic Arms Limitation Talks.* New York: Doubleday, 1980.

Stern, Philip M., with the collaboration of Harold P. Green. *The Oppenheimer Case: Security on Trial.* New York, Harper & Row, 1969.

Sumner, William G. *War and Other Essays.* Freeport, NY: Books for Libraries Press, 1970.

Sun Tzu. *The Art of War,* trans. Samuel B. Griffith. Oxford: Oxford University Press, 1963.

Sykes, Christopher. *Orde Wingate: A Biography.* Cleveland: World Publishing, 1959.

Tacitus. *Complete Works of Tacitus,* ed. Moses Hadas and trans. Alfred John Church and William Jackson Brodribb. New York: Random House, Modern Library College Edition, 1942.

Talbott, Strobe. *Deadly Gambits: The Reagan Administration and the Stalemate in Nuclear Arms Controls.* New York: Alfred A. Knopf, 1984.

Taylor, A. J. P. *How Wars Begin.* London: Hamish Hamilton, 1977, 1979.

———. *Politics in Wartime and Other Essays.* London: Hamish Hamilton, 1964.

Bibliography

Temperley, Major-General A. C. *The Whispering Gallery of Europe.* London: Collins, 1938.

Thomas, Gordon, and Max Morgan Witts. *Enola Gay.* New York: Stein and Day, 1977.

Tocqueville, Alexis de. *Democracy in America,* Vol. I (1835), ed. J. P. Mayer and trans. George Lawrence. New York: Harper & Row, 1966.

Toland, John. *No Man's Land: 1918, The Last Year of the Great War.* New York: Doubleday, 1980.

Toynbee, Arnold J. *Civilization on Trial.* New York: Oxford University Press, 1948.

————. *War and Civilization,* selected by Albert Vann Fowler from *A Study of History.* New York: Oxford University Press, 1950.

Trewhitt, Henry L. *McNamara: His Ordeal in the Pentagon.* New York: Harper & Row, 1971.

Truman, Harry S. *Off the Record: The Private Papers of Harry S. Truman,* ed. Robert H. Ferrell. New York: Harper & Row, 1980.

Tuchman, Barbara W. *The Guns of August.* New York: Macmillan, 1962.

————. *The Proud Tower: A Portrait of the World Before the War: 1890–1914.* New York: Macmillan, 1966.

Ulam, S. M. *Adventures of a Mathematician.* New York: Charles Scribner's Sons, 1976.

Villari, Pasquale. *The Barbarian Invasions of Italy,* trans. Linda Villari. New York: Charles Scribner's Sons, 1902.

Wallace, D. Mackenzie (Sir Donald). *Russia* (1877). New York: Reprinted by AMS Press, 1970.

Walters, Vernon A. *Silent Missions.* New York: Doubleday, 1978.

Wellard, James. *Babylon.* New York: Saturday Review Press, 1972.

Whitten, Lieut.-Colonel F. E. *The Decisive Battles of Modern Times.* London: Constable, 1913.

Bibliography

Williams, Beryl, and Samuel Epstein. *The Rocket Pioneers: On the Road to Space.* New York: Julian Messner, 1955.

Williams, Francis. *Twilight of Empire: Memoirs of Prime Minister Clement Attlee.* New York: A. S. Barnes.

Wilson, Andrew. *The Bomb and the Computer: Wargaming from Ancient Chinese Mapboard to Atomic Computer.* New York: Delacourte Press, 1968.

Wilson, Edmund. *Red, Black, Blond and Olive: Studies in Four Civilizations: Zuñi, Haiti, Soviet Russia, Israel.* New York: Oxford University Press, 1956.

York, Herbert. *Race to Oblivion: A Participant's View of the Arms Race.* New York: Simon and Schuster, 1970.

Zuckerman, Solly. *From Apes to Warlords.* New York: Harper & Row, 1978.

INDEX

Index

181

Index

Index

Index

NATO (North Atlantic Treaty Organization), 43, 51, 95, 135
Natural History (Pliny the Elder), 35
nature, control over, 117–118, 142
Nazi regime, 93
necks, saving of, 42
Neumann, John von, 28, 30, 31–32
neutrality, as obsolete, 51
neutron bomb, 86
New Mexico, atomic test in, 23
New York Times, 24, 28
New York Tribune, 62
Nicholas II, tsar of Russia, 76
Nitze, Paul, 87, 131
Nixon, Richard, 44, 57, 134, 135
Nobel Prize, 97
Noel-Baker, Phillip, 82
North Atlantic Treaty Organization (NATO), 43, 51, 95, 135
North Vietnam, bombing of, 135
nuclear deterrence, 9, 29, 86, 131
nuclear superiority, 57
nuclear targeting, 72
nuclear weapons, 11, 18, 31, 33, 86, 87, 135. *See also* atomic bomb; thermonuclear bomb

Odom, William, 136
Office of Scientific Research and Development, 39
Ofstie, Ralph A., 50
"On Fighting a Nuclear War" (Howard), 143
On Thermonuclear War (Kahn), 54
On War (Clausewitz), 107, 138
Oppenheimer, J. Robert, 21, 22, 27–28, 29, 30, 66, 78, 79, 80, 81, 83, 99
Oxnam, G. Bromley, 26–27

Pais, Abraham, 79–80
Palmerston, Lord, 115, 123
Passionate Prodigality, A (Chapman), 127
Patriarch of Constantinople, 137
Paul, prince of Yugoslavia, 117
Pauli, Wolfgang, 129
Pauling, Linus, 33
peace, 71, 79, 89, 104, 105, 106, 111, 120, 122–123, 127–128
Peace and War (Aron), 118
Peace and War (Ferrero), 117
Perle, Richard, 88
Pershing, John "Black Jack," 128
Persian Gulf, war over, 143
Persian Letters (Montesquieu), 36
Peter the Great, tsar of Russia, 59
Philippine insurrection, 122
Phillips, W. Alison, 122–123
physics, 84, 97, 98
Pickering, William, 41
Planck, Max, 98
Plehve, V. K., 115
Pliny the Elder, 35
Plutarch, 36, 104
Polaris program, 42
Politburo, 40
Poltava, 59–60
Porto, Luigi da, 89
Poseidon missiles, 56
Potsdam Conference (1945), 22, 23
Power, Thomas S., 51, 52, 53, 63–64, 118
pre-emptive attack, 53
Private Papers of Henry Ryecroft, The (Gissing), 97
psywar, 119
Pultowa, Battle of, 62

Quincy, Josiah, 106
Quintus Curtius, 103
Qumran community, 103

185

Index

Index

187

Index

188

ABOUT THE AUTHORS

Thomas Powers won a Pulitzer Prize in 1971 for a series of articles on Diana Oughton, a member of the Weathermen. He is the author of several books, among them *The Man Who Kept the Secrets*. He is a contributing editor of *The Atlantic* and on the board of *Nuclear Times*. Mr. Powers is also on the advisory board of the Nuclear Weapons Databook Project and a member of the Council on Foreign Relations. He lives in Vermont.

Ruthven Tremain is well known as the creator of *The Calendar for Children* (Macmillan, 1963, 1965–1976). She has also written many books, among them *Fooling Around with Words* (Greenwillow, 1976). Her latest book is *The Animals' Who's Who*. She lives in New York City.